Weekend Mechanic's Guide to ENGINE REBUILDING

Weekend Mechanic's Guide to ENGINE REBUILDING

Paul Dempsey

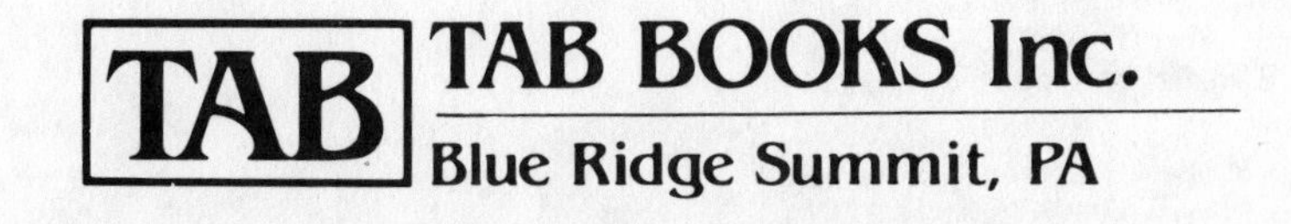

FIRST EDITION
FIRST PRINTING

Library of Congress Cataloging in Publication Data

Dempsey, Paul.
Weekend mechanic's guide to engine rebuilding / by Paul Dempsey.
p. cm.
Includes index.
ISBN 0-8306-9117-0 ISBN 0-8306-3117-8 (pbk.)
1. Automobiles—Motors—Maintenance and repair. I. Title.
TL210.D44 1989
629.2'504—dc19 88-32999
CIP

TAB BOOKS Inc. offers software for sale. For information and a catalog, please contact TAB Software Department, Blue Ridge Summit, PA 17294-0850.

Questions regarding the content of this book should be addressed to:

Reader Inquiry Branch
TAB BOOKS Inc.
Blue Ridge Summit, PA 17294-0214

Kimberly Tabor: Acquisitions Editor
Peter D. Sandler: Technical Editor

Contents

Acknowledgments

I want to thank Gary Fish, who has written seven technical publications and is manager and machine-shop foreman at Quade Auto Supply, Inc., in Houston. Gary always took time to answer my questions and has the experience and knowledge to make his answers meaningful.

Introduction

This book is written for the do-it-yourselfer who wants to rebuild an auto engine and do the job right. You will learn how to save money on parts, how to select a reliable machinist and how to check his work. Most importantly, you will learn how to assemble that engine for 100,000 miles or more of trouble-free service. Service bulletin updates for most domestic and many foreign engines are included, together with information that you cannot find in the factory service manuals.

A rebuilt engine should have the durability of a new engine, and, because the cylinders usually have to be bored oversize, it might even be slightly more powerful than the day it left the showroom floor. The cost of a rebuilt engine is nominal when compared to the price of a new car or a factory-built engine, and the workmanship should at least equal anything coming out of Detroit or Yokohama.

Nor is the work complicated, if taken a step at a time. This book describes the whole process: troubleshooting, organizing the workspace, making notes and labeling the electrical and vacuum connections, disassembling to discover what went wrong and why, machining operations, and finally assembling the engine in accord with factory standards.

Of course, nobody would rebuild an engine unless there were practical justification for it. But engine rebuilding is also fun—a highly creative, productive, challenging activity.

1 CHAPTER

Engine Wear

SOMETIMES THE DECISION TO PERFORM MAJOR ENGINE REPAIRS CAN BE AS PLAIN and as unambiguous as a connecting rod through the side of the block. At other times, it is a judgment call and depends upon the worth of the vehicle, the length of time you expect to keep it, the extent of the damage, and the estimated cost of repair.

In the absence of death rattles, puddles of oil on the garage floor, or other dramatic symptoms, engines with 100,000 miles or more on the clock are potential candidates for repair. A leakdown test (described in this chapter) should settle the matter. Oil burning—as opposed to oil leaks—is another leading indicator. In general, an engine that burns 1 quart of oil every 300 or 400 miles is in need of serious work.

HOW ENGINES WEAR

In the most fundamental sense, all engines wear out in the same way. A thin film of oil separates the piston rings from the cylinder walls, but this oil film becomes marginal each time the piston reverses direction at the top and bottom of the stroke. Some metal-to-metal contact occurs.

Cylinder Bore

In this process, both the rings and cylinder walls wear. Combustion gases blow by the rings, contaminating the oil supply and accelerating wear on other components. Engineers calculate that the best that can be expected from any reciprocating engine is one billion revolutions, regardless of engine speed, bore diameter, or quality of fit and materials.

For a big V-8, one billion revolutions is roughly equivalent to 200,000 miles of moderate stop-and-go driving. Of course, very few engines make it this far: a more typical figure is between 80,000 and 120,000 miles. Reality falls disappointingly short of theory. To understand why, it is necessary to look more closely to each of the major engine components.

Compression. The most critical aspect of any engine is the sliding seal formed between the compression ring ODs and the cylinder bore (Fig. 1-1). Gasoline vapor and air must be compressed before they will explode. The gases generated by that explosion must be contained above the piston where they will perform useful work. The rings have some residual tension, which is why they are difficult to install in the first place, but the basic sealing mechanism is illustrated in Fig. 1-2. Gas pressure against the back of the ring forces the

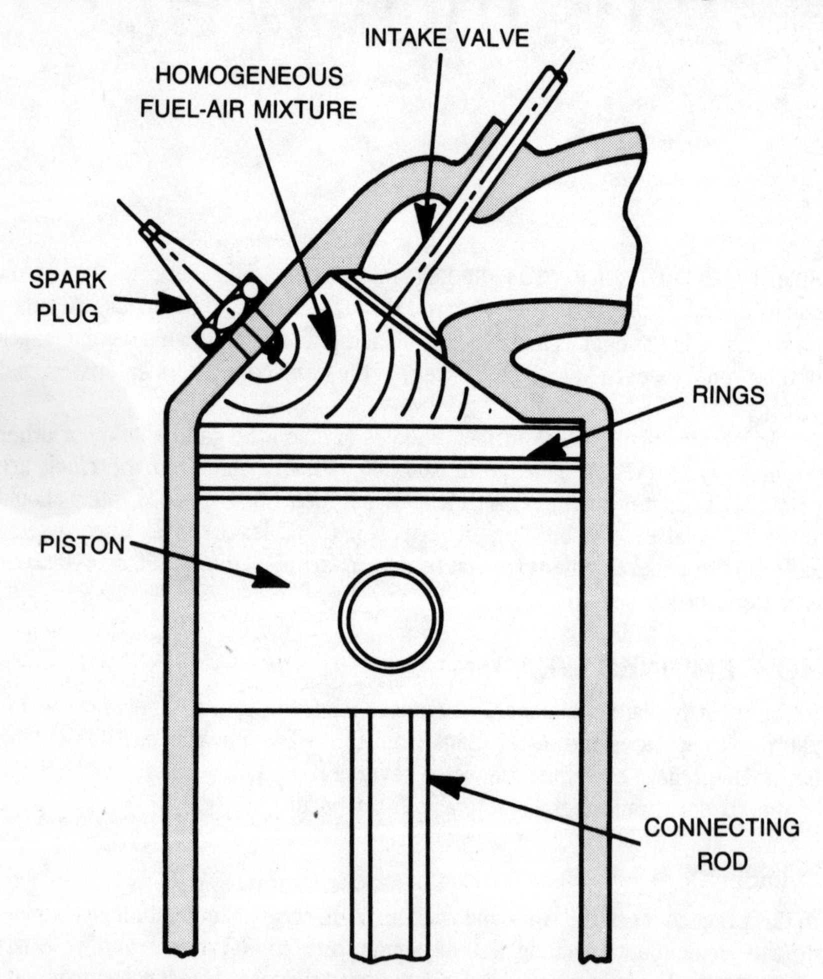

Fig. 1-1. This is where it all begins. Air and fuel mixture is admitted into the combustion chamber by the intake valve, compressed by the piston, and ignited. The explosion pushes the piston downward in the power stroke. Piston rings function as seals, containing the gas pressure to prevent power losses. (Courtesy Technology Today)

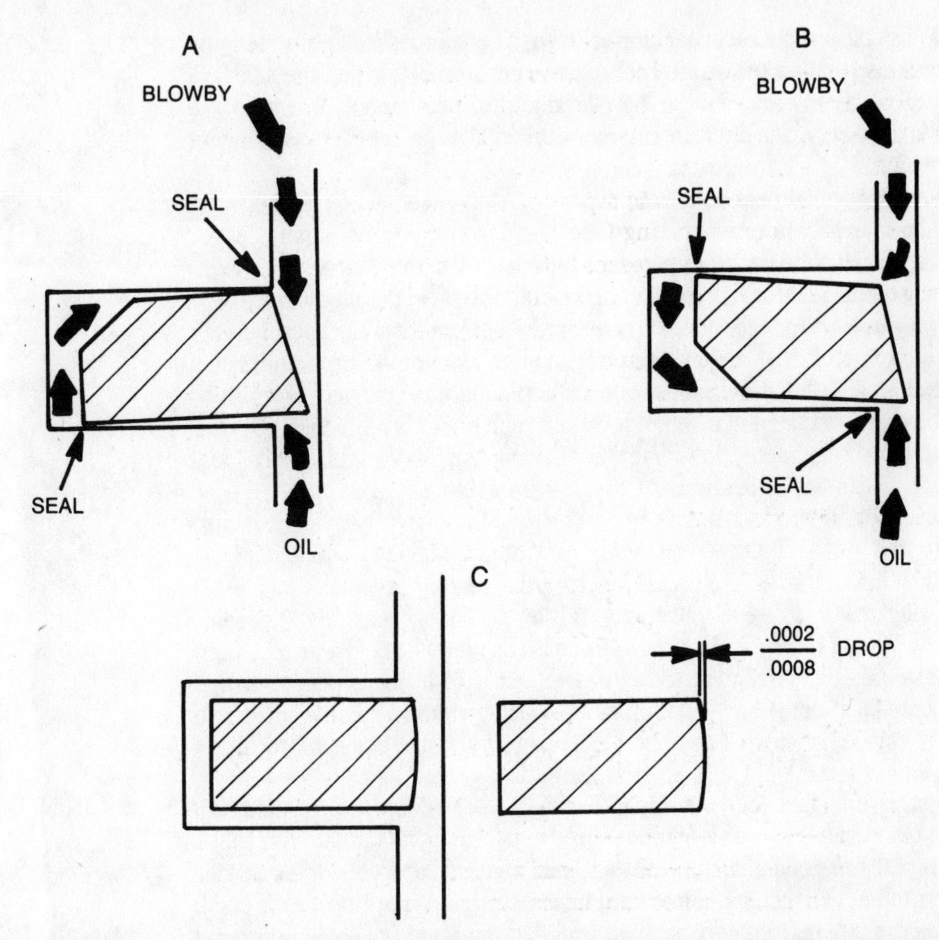

Fig. 1-2. All compression rings function as dynamic seals: gas pressure forces the rings outward against the cylinder bore. The greater the pressure, the more tightly the rings hug the bore. Modern ring designs also address the question of leakage between the ring and piston. Drawing "A" illustrates a tapered face positive twist ring. When installed in the engine, the ring tilts about its axis, sealing the piston grooves against blow-by from above and oil from below. Drawing "B" shows a later evolution of this ring. Now the bevel is on the lower inner edge, and ring twist is reversed. This design gives better oil control, during coastdown, when the throttle is closed and cylinder bores are subjected to high vacuum. Some blow-by control is sacrificed. At wide-open throttle both the positive-twist ring "A" and the reverse-twist ring "B" lie flat in their grooves. The radius-face, or barrel-face, compression ring, shown at "C," has become popular for high performance applications. A slight radius on the ring OD gives high contact pressures against the bore and maintains this contact when the piston reverses direction at the end of each stroke. (Courtesy Sealed Power Corp.)

ring face into firm contact with the cylinder bore. The greater the pressure above the piston, the tighter the rings seal.

Rings also act as oil seals restricting the amount of crankcase oil that enters the combustion chamber. The lower portion of the bore is wetted with oil thrown off the crankshaft journals. Bucketlike serrations in the oil ring fill on the down stroke and distribute the oil along the length of the bore. The surplus is wiped off by the second compression, or scraper, ring. Only a very thin film of oil remains to lubricate the No. 1 compression ring, which helps to explain why this ring tends to wear more than the others.

This double seal, segregating combustion processes from engine mechanicals, is what makes an internal combustion possible. When the seal fails because of wear, raw gasoline and by-products of combustion leak into the crankcase. The oil, which circulates to every friction surface, becomes an agent of destruction.

While marginal lubrication during the two piston reversals on each stroke sets an upper limit to bore wear, there are other factors. Some engines have, as it were, the genetic background that predisposes them to long lives. Thin-wall casting techniques, which reduce the amount of iron in the block, have now become universal. Although these engines are fuel-efficient simply because they weigh less, some of them are prone to rapid wear. Part of the problem seems to be associated with the breakdown of multigrade lubricants; part of it has to do with block distortion; and much of it is the result of soft iron, which is easier to pour at close tolerances. Japanese, most European, some Oldsmobile, and all Cadillac blocks are exceptions to the soft-iron rule.

Wear is not only a function of the hardness of the block material; it is also related to its rigidity. To my knowledge, all modern engines come out of the factory with built-in stresses, created during the casting process. (In the old days, quality builders like Cadillac and Harley-Davidson seasoned "green" castings by storing them for six months or so in the weather before machining.) Casting stresses eventually work themselves out and, in the process, distort cylinder bores and upset bearing alignment. This is why racers often prefer to build their engines from well-used and dimensionally stable blocks. It is also a good argument for rebuilding a worn engine.

The same principle does not apply to cylinder heads, which are subject to high temperatures that go beyond the transformation range. At these temperatures, the grain structure of the metal changes. Iron becomes harder and prone to develop cracks. Aluminum heads are even more prone to crack failure and can suffer from corrosion as well. The insignificant-looking ground straps between the head and block or head and fire wall must be in place if galvanic corrosion is to be eliminated.

Casting stresses can be relieved, although most machinists would be surprised by the request. Stress relieving involves heating the iron head and block to about 100 degrees F. below the transformational range and allowing the parts to cool slowly. The heat realigns the grain structure, stabilizing the parts. Stress relieving would be indicated on a particularly valuable engine that overheated and was quenched with cold water.

Thermal distortion affects cylinder bores and crankshaft bearing alignment. In Chapter 5 we discuss a special kind of thermal distortion that bends aluminum cylinder heads used with iron blocks. Some cylinder bore distortion develops as the engine reaches operating temperature and must be more pronounced on those engines with siamesed cylinders. The center cylinders on each bank of the Chevrolet 400 engine are *siamesed*; that is, their adjacent edges fill the area normally reserved for the water jacket. These edges, flanked by iron, must run hotter than the water-cooled portions of the bore. MG and Jaguar also use this method of construction.

Fasteners—the nuts, bolts, and studs that hold the engine together—are another cause of distortion and premature wear. Once your engine is apart, you can demonstrate this by miking the bores, from the pan side with the head(s) installed and torqued. Repeat the measurements with the heads removed. Head bolt-induced distortion can also affect the valve seats, although that is harder to measure.

To some extent, fastener distortion can be compensated for during the rebuilding process. Cylinder bores should be miked with the main bearing caps torqued down and the caps should remain installed during boring and finish-honing. Most good machinists will use a torque plate—a kind of dummy cylinder head with holes corresponding to the cylinder bores—during cylinder machining.

Factory tolerances can work for or against durability. For example, Ford factory 351W specifications call for a piston-to-cylinder bore clearance of 0.0018 to 0.0026 in. Although the tolerance is small in absolute terms—only 0.0008 in.—it means that ''loose'' pistons have percent more clearance than their ''tight'' cousins. Ford is not unique in this regard: other manufacturers allow themselves similar latitude.

Loss of compression is a fairly subtle symptom, and so long as all cylinders are equally affected, few drivers would notice anything wrong. Power and fuel economy will be off ever so slightly. Loose pistons may rattle and slap, but the noise usually goes away as the engine warms (Fig. 1-3).

What does get attention is the loss of the oil seal function. The engine will consume increasing quantities of lube oil, spark plugs may foul, and older vehicles will trail blue smoke out the exhaust. Newer vehicles, equipped with catalytic converters, will retain a sanitary exhaust until the converter overheats and fails.

Tests. There are three commonly used tests that measure compression or some approximation of it. In order of ascending accuracy, they are:

Dynamic Compression Test: Not particularly accurate, but fast. With a tachometer hooked up to the ignition system, run the engine until it reaches stable operating temperature. Short out one spark plug lead at a time, recording the rpm drop. A weak cylinder obviously works less than its stronger brethren and its loss has less effect on engine rpm. This test should not be performed on certain engines with electronic ignition systems and GM warns against the test on vehicles with catalytic converters.

Cranking Compression Test: Favored by most mechanics, this test measures compression directly (Fig. 1-4). Remove all the spark plugs, marking the wires as necessary, and disable the ignition system by disconnecting the small-diameter wire going to the negative side of the coil. (The old method of grounding the coil-to-distributor-cap wire can damage electronic ignition systems.) If a remote starter switch is used, be sure to turn the ignition key to the ''run'' position. Block the throttle and the choke butterflies full open.

Gauge each cylinder, recording the compression reading on the four compression stroke. While cranking compression specs are sometimes published, the numbers are not terribly meaningful. Look instead for consistency: the weakest cylinder should be within 75 percent of the highest. Low compression on two adjacent cylinders points to a blown head gasket. If compression exceeds

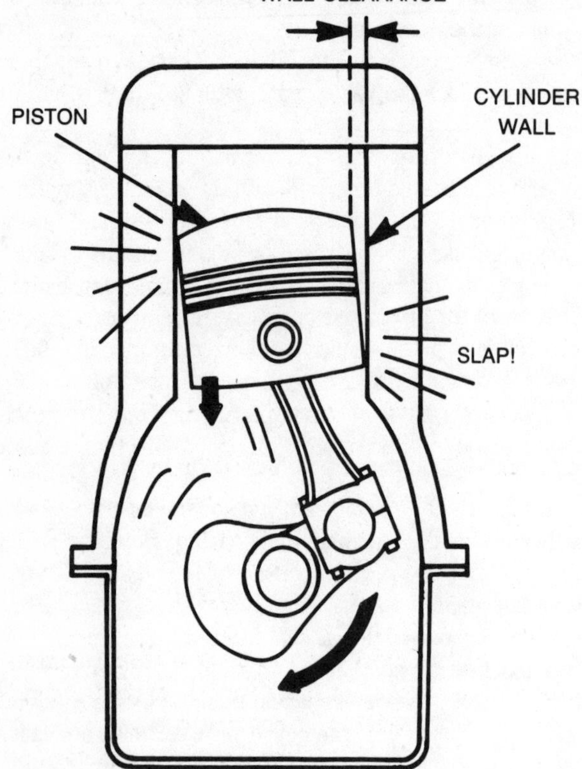

Fig. 1-3. Piston slap, heard as a hollow, rpm-sensitive knock on a cold engine, means that the pistons are tilting in the bores.

Fig. 1-4. Making a compression test with a FoMoCo combination gauge and remote starter switch. All spark plugs must be removed for the test to approach validity. Electronic ignition systems must be disconnected—not merely grounded—and, if a remote starter switch is used, the ignition key should be turned to the "run" position. Some ignition switches go to ground "off" position, creating a dead short when a remote starter switch is used.

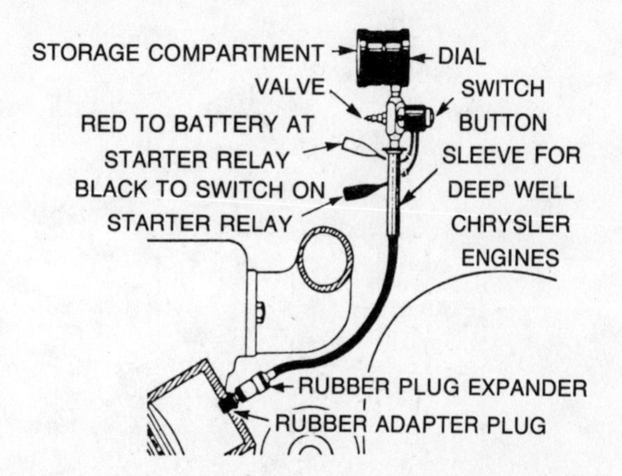

Fig. 1-5. A K-D combination compression and leakdown tester.

the original specifications, suspect heavy carbon buildup in the combustion chambers.

Squirt about 30 cc (two tablespoons) of oil into each cylinder through the spark plug ports. Spin the engine a few revolutions to distribute the oil and make a second compression test. If the "wet" test results in significantly higher compression, the problem is associated with the lower engine (piston rings, pistons, cylinder bores). If low cylinders fail to recover compression, the fault is in the valves and/or cylinder head gasket.

Leakdown Test: By far the most definitive test, but requires some delicate fiddling (the crankshaft must be turned to bring each piston to exact top dead center). Compressed air is introduced to each cylinder through a special fitting that replaces the spark plug (Fig. 1-5). A gauge registers the percentage of leakdown, which for a mechanically perfect cylinder can be as low as 2 percent and which should in no event exceed 25 percent. The point of exit of escaping air indicates the source of the leak. For example, air hissing through the PCV valve or oil-filter fitting must have passed by the rings, and air from the carburetor inlet had to pass the intake valve, which of course is closed at tdc. An exhaust valve leak will show up at the tailpipe; a blown head gasket may aerate the coolant in the radiator, or it may leak through an adjacent cylinder or at the head/block interface.

Bearings

From one point of view, an engine is nothing more than a collection of bearing surfaces. When a mechanic talks about "bearings" he means those bearings associated with the crankshaft, that is, the main bearings (which support the crank in the block) and the crankpin bearings (which support the connecting rods on the crankshaft pins). There is good reason for this.

Wear on the crankshaft journals and bearings goes hand-in-hand with cylinder wear and, in one sense, is even more critical. Barring other complications, cylinders continue to wear until the engine no longer develops enough compression to start; but a seized crankpin (connecting rod) bearing is a catastrophic event, often resulting in a connecting rod through the side of the block.

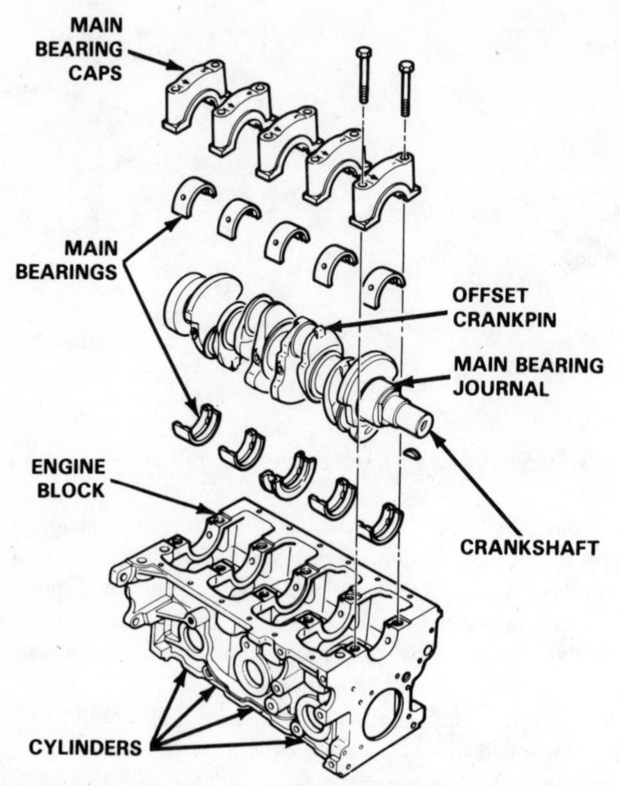

Fig. 1-6. Four-cylinder engine block, showing crankshaft and main bearings. Flanges on the center main bearing upper insert take out crankshaft thrust loads, generated by an automatic transmission or when the clutch is depressed on a vehicle with a manual transmission.

Figure 1-6 shows the internals of a typical four-cylinder engine. Five main bearings—located at each end of the crank and between each of the crankpins, or throws, support the shaft in the block. The bearings consist of precision insert-type shells, mounted in split bosses, or saddles. Flanges on one of the center main bearing inserts take out fore-and-aft thrust forces.

The connecting rods pivot on crankpin bearings, which are smaller versions of the main bearings.

Figures 1-7 and 1-8 illustrate the basic lubrication circuit. The pump lifts oil from the oil pan, or sump, and delivers it under pressure to the bearings and rocker arms. Oil returns to the sump by gravity. An oil galley, running the length of the block, distributes oil to the camshaft and crank bearings through drilled passages (Fig. 1-9). The crankshaft is a kind of rotating gallery: oil enters the hollow crank at the center main bearing saddle and is distributed to each of the crankshaft bearings (Fig. 1-10). Some camshafts work the same way; others are lubricated by an external circuit.

Once the engine starts and oil pressure comes up, journal bearings float, somewhat off center, on a film of oil, as shown in Fig. 1-11. Wear and friction

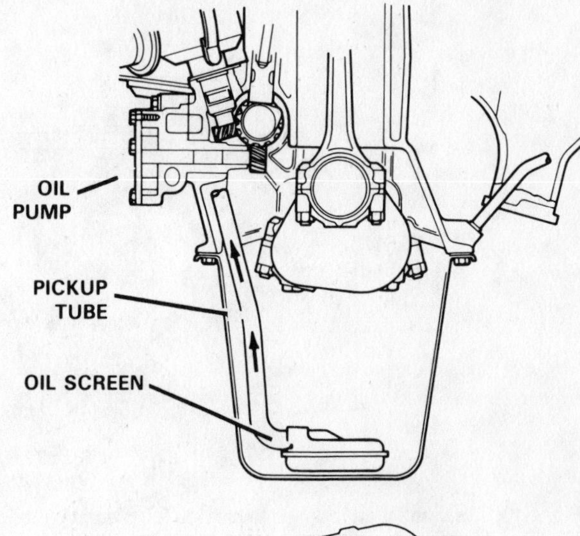

Fig. 1-7. The oil pump draws oil from the oil pan through a pickup tube and screen. A clogged screen or air leaks on the suction side of the pump can reduce oil pressure to dangerous levels. (Courtesy Federal Mogul Corp.)

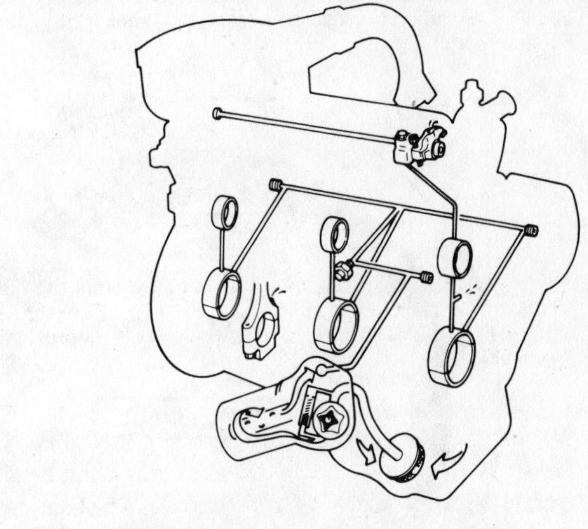

Fig. 1-8. The pump sends pressurized oil through the filter, then to the crankshaft, camshaft, and main oil gallery. (Courtesy Federal Mogul Corp.)

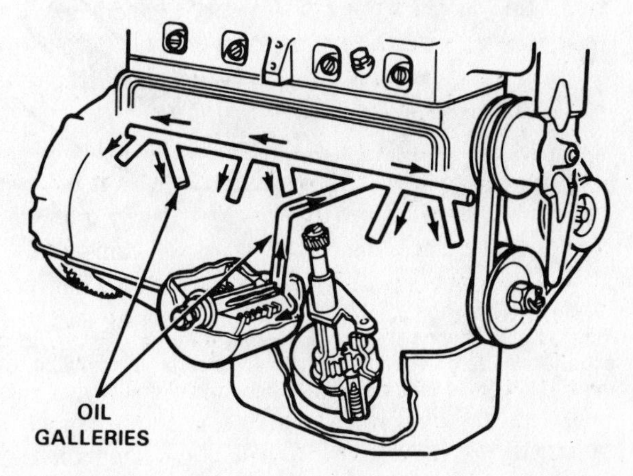

Fig. 1-9. Drilled passageways interconnect with the main oil gallery to bring lubrication to the various bearing surfaces. (Courtesy Federal Mogul Corp.)

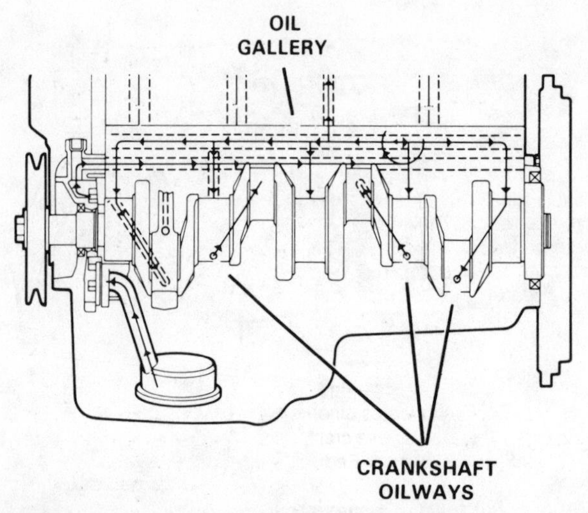

Fig. 1-10. The crankshaft is drilled to provide full pressure lubrication to main and crankpin bearings. Standard practice is to supply the crank via the main bearing saddles. (Courtesy Federal Mogul Corp.)

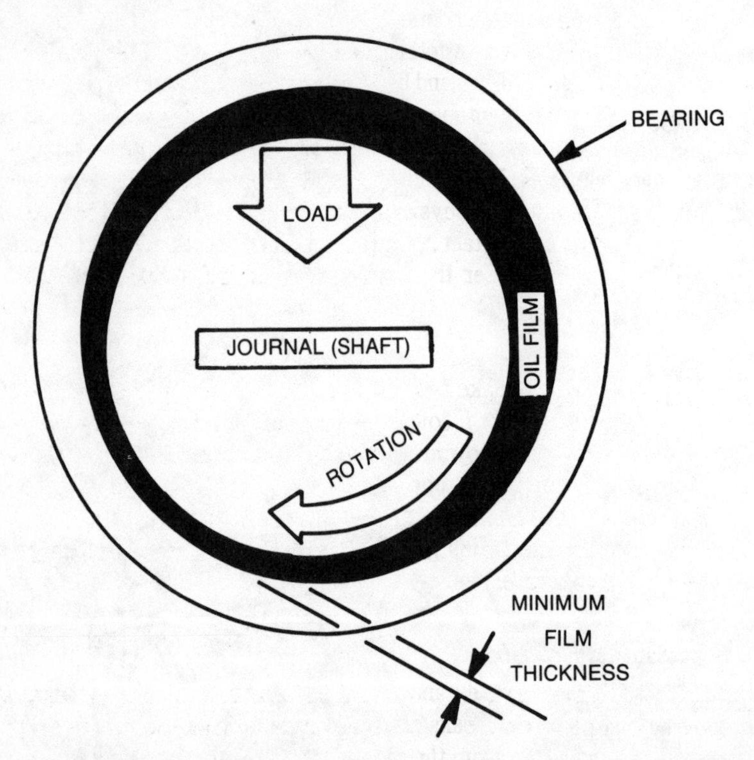

Fig. 1-11. In theory, a journal bearing rides on a cushion of oil and, once underway, experiences zero wear. In practice, some wear is inevitable. Bearing clearances increase, the film becomes progressively thinner and eventually fails. (Courtesy Technology Today)

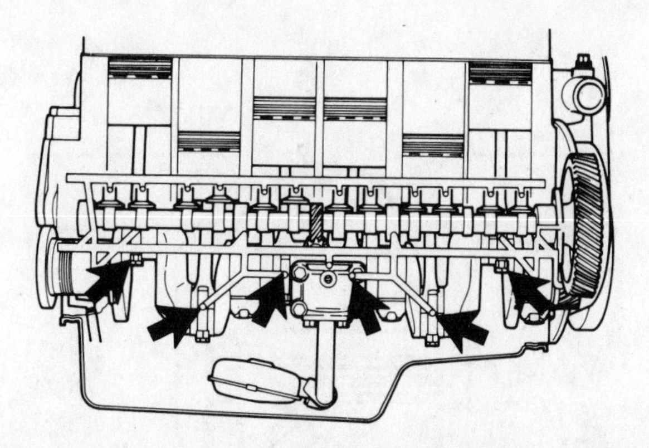

Fig. 1-12. Crankshaft throw-off lubricates other engine components at a rate which is primarily determined by bearing clearances. Loose crank or camshaft bearings flood the cylinder bores, increasing oil consumption. (Courtesy Federal Mogul Corp.)

are, under these circumstances, nonexistent. If oil pressure is maintained, crank and cam bearings do not touch their journals and cannot wear.

Note that individual crankshaft bearings are not sealed: oil leaks past the edges of the bearing and must be continually replenished by the pump (Fig. 1-12). The rate of bearing leakdown is determined by oil viscosity—the heavier the oil, the more persistent the film—and by the clearance between the bearing and its journal. Some wear is inevitable (even for a theoretically perfect bearing) and after many miles, bearing clearances open, increasing the rate of leakdown. The oil film grows progressively thinner. Eventually it can no longer support the load, and the journal makes physical contact with the bearing. The heat developed by metal-to-metal contact carburizes what oil remains into an abrasive goop. A few revolutations later the bearing welds itself to the journal.

Noise

Dangerously loose bearings make a characteristic knock, depending upon bearing location and engine rpm. The following signs help identify the bearing:

Rod Bearing: Originating low in the cylinder block and in phase with engine rpm, rod bearing knock usually stops when the spark plug wire to the affected cylinder is grounded (Fig. 1-13).

Main Bearing: Loose mains generally produce a dull thud, which grows more insistent as load increases.

OIL CONSUMPTION

Oil consumption and bearing clearance are closely related. The crankshaft acts as an impeller, flinging oil that leaks past the journal bearings on to the cylinder walls. Leakdown—primarily from the crank, but also from the camshaft—is the source of cylinder lubrication. Uncontrolled leakdown floods the cylinders, collapsing scraper rings, and allowing oil to enter the combustion chambers.

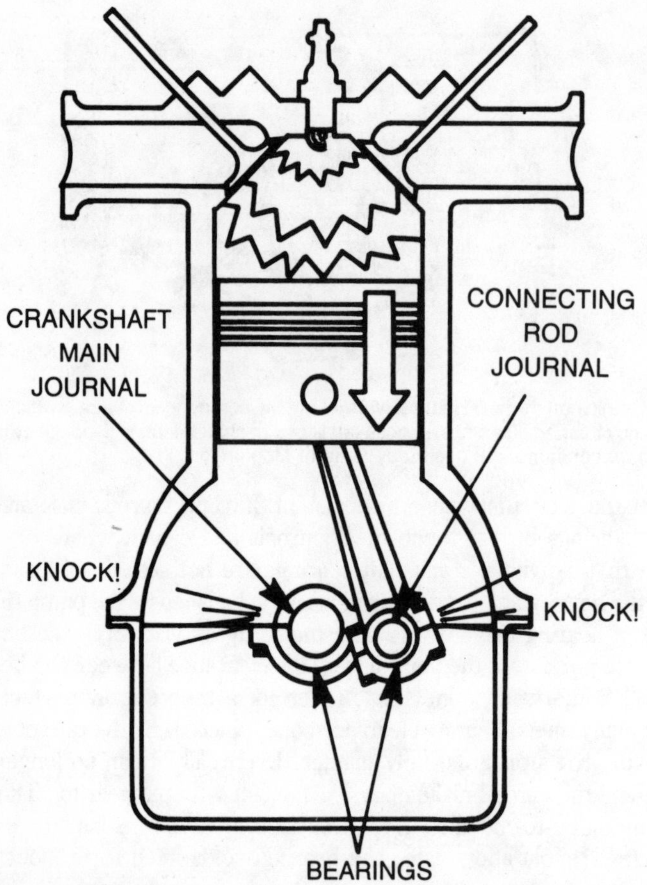

Fig. 1-13. Rod knocks can be described as a kind of tapping that increases in frequency with engine speed. Main bearing knocks are generally deeper, more hollow sounding, and often disappear under light loads.

Severely worn engines use oil, but this does not mean that engine condition can be intuited from the dipstick. You need to determine how and why the oil is consumed.

Oil Burning

Persistent oil burning—characterized by a trail of blue smoke out the exhaust (on pre-pollution engines) and by heavy, black carbon deposits on the spark plugs—is the classic symptom of cylinder bore/piston ring wear. Verify with a compression or leakdown test.

It should also be remembered that any oil-wetted area subject to manifold vacuum can be a source of oil entry into the combustion chambers. For example, a leaking intake manifold gasket on V-8 engines that use the manifold to roof over the cylinder valley can dramatically increase oil consumption. Failed turbocharger oil seals have the same effect. Positive crankcase ventilation (PCV) systems vent the crankcase to the intake manifold, as illustrated in Chapter 4

(Fig. 4-9). Crankcase air leaks, caused by a failed rocker arm cover gasket or crankshaft oil seal, or occasionally by something as simple as an improperly seated dipstick, can reverse PCV system flow and might allow oil to enter the manifold. A loose or missing PCV baffle can have the same effect. The baffle is always located below the PCV pickup connection, e.g., under the intake manifold on Ford 302 engines or inside a rocker arm cover. Test by running the engine with the PCV system disconnected and plugged.

Three-valve Honda CCVC engines are a special case. These engines employ O-ring seals around the smaller intake valve bodies. The O-rings fret and eventually fail, allowing large amounts of oil to migrate from the cylinder head into the combustion chambers.

Intermittent oil burning, signaled by a puff of smoke during start-up, means that the intake valve seals and possibly valve guides have failed. Loose exhaust valve guides can contribute to overall oil consumption. Chevrolet V-8s tend to flood their cylinder heads and are very sensitive to valve seal problems.

Leaks

It has been calculated that an engine which loses one drop of oil every 150 ft. will consume a quart in 500 miles.

Most common sites of oil leaks are:

- ☐ Rocker arm cover gasket
- ☐ Oil pan gasket
- ☐ Timing chain cover gasket
- ☐ Intake manifold end seals (V-8 engines)
- ☐ External accessory mounting gaskets, such as fuel pump, oil pump, oil filter
- ☐ Rear and front crankshaft oil seals
- ☐ Camshaft, oil gallery plugs

Note that crankshaft oil seals are well down on the list. Oil leaks from the rear or front of the engine are not *prima facie* evidence of seal leakage; in the great majority of cases the fault is elsewhere.

Any leak will be intensified by positive crankcase pressure, developed because of a faulty ventilation system (sticking valve or clogged hoses), excessive blowby (worn rings and/or cylinder bores), or excessive ignition advance. More oil will be lost if the engine is lugged or operated at high speed.

Locating the source of a small leak can be frustrating—oil migrates in response to gravity and wind currents from the fan and the forward motion of the vehicle. In some cases, the vehicle may have to be tilted to reproduce driving conditions.

Begin by cleaning the external surfaces of the engine. Run the engine to operating temperature and, possibly using a mirror, search for the seepage, staying well clear of the fan and drive belts. If you have to get under the vehicle, support it, wheels up, on stable jackstands.

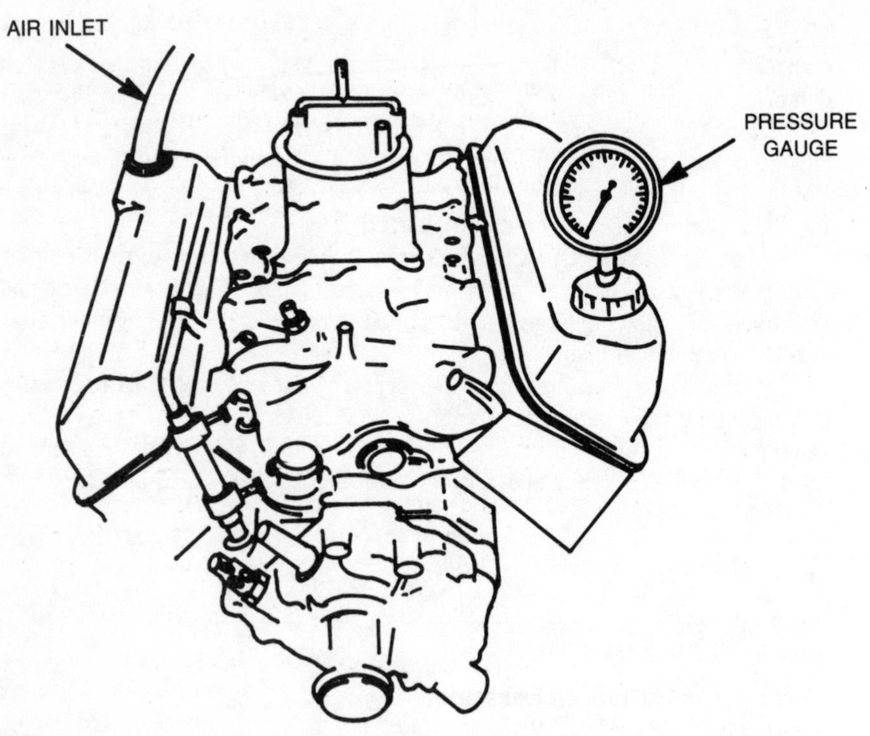

Fig. 1-14. Hard-to-find oil leaks can be ferreted out by pressurizing the crankcase.

An alternative procedure is to pressurize the crankcase (Fig. 1-14). Connect an air line to the oil filler opening or dipstick tube, regulated at no more than 5 psi (pounds per square inch). Soap the engine down and look for bubbles. Light foaming around the rocker arm cover gaskets is permissible. If this test fails to produce definitive results, the transmission and flywheel can be removed for access to the rear oil seal, rear main bearing parting line (i.e., the main bearing web and cap interface), and oil gallery plugs at the rear of the block. Rope crankshaft seals will pass some air and remain serviceable; newer lip seals may exhibit foaming.

CHOICES

There are several ways to deal with a sick engine. One alternative is to sidestep (or at least put off) the problem by purchasing a used engine from a wrecking yard.

Prices vary with mileage and demand. In the Houston area, Chevrolet 454s range between $300 for a rebuildable hulk to more than $1500 for low-mileage examples. Obsolete, but still desirable, Ford Clevelands and pre-Electronic Fuel Injection Chevy 350s go for a premium. One can sometimes find a bargain, but don't count upon it. Most yards are linked by a voice-over telephone line which tends to standardize prices. Some yards also subscribe to a computerized quoting service, such as the Orion Network. Orion currently has 600 subscribers on the West Coast.

Used engines from Japan are available in metropolitan areas. One source is Japanese Motors International, Inc. (638 Pickering, Houston, TX 77091). While prices fluctuate with the yen/dollar exchange rate, a Toyota 20R engine with 30,000 to 40,000 miles on the clock currently goes for less than $400, or about half the cost of a rebuild. In addition, owners who are willing to risk the wrath of the Environmental Protection Agency can purchase high performance engines that were never sold in this country. For example, the 18 R-G, a double-overhead-cam, 160-hp alternative to the anemic 20R, sells for $520. None of these engines has more than 40,000 miles on it because Japanese insurance and inspection laws make it impractical to keep a vehicle more than two or three years. Engines are warranted and the company supports them with parts sales.

Normally, the engine selected should be identical or closely interchangeable with the original, with the same attachment hardware, throttle, hose, and exhaust connections. Emission controls and computerized engine management systems must be entirely compatible. Listen to the engine run and, if possible, make a compression or leakdown test prior to purchase. An oil analysis may also be appropriate: R.W. Strecker, Texaco, Inc., 2000 Westchester Ave., White Plains, NY 10650 can provide the details.

Most wrecking yards certify the mileage and provide a 30- or 90-day exchange-only warranty. Some paperwork may be required to keep the vehicle title current.

Another approach is to purchase a factory rebuilt engine. These engines are built on an assembly line (hence the term "factory" rebuilt) and are marketed through auto dealerships, auto parts stores, or directly from the remanufacturer. Most are sold as "long blocks," that is, as complete engines from oil pan to valve covers. All that is missing are manifolds and accessories, such as the starter motor and alternator. The same package, less cylinder head(s), is known as a "short" block. If you decide for some reason to keep your old castings, you will be assessed a core charge. Most rebuilders offer a 90-day warranty, good for an exchange engine.

Some of these engines, particularly those certified by the original equipment manufacturer (OEM), are quality products. Others are ruthlessly built down to a price and cobbled together from inferior parts by unskilled labor.

Another alternative is to turn the job over to a professional mechanic. The cost of such an exercise is high, but, by the same token, the mechanic should be willing to stand by his work. A secondary advantage is that you keep your original engine castings.

The most creative and satisfying approach is to do the work of disassembly and assembly yourself, farming out the necessary machine work. This approach gives you complete control over the process—you can specify the quality of replacement parts and check the accuracy of machining operations while the engine is still apart and before a mistake becomes catastrophic. It is also cost-effective, provided that leisure time is not expensed as a labor charge.

SCOPE OF WORK

An *overhaul* is a partial rebuild that is appropriate for a moderately worn engine. It entails a valve job, rings, and new standard-sized main and crankpin bearings. An engineered set of piston rings (which exert greater side pressure than conventional, or factory, ring sets) can compensate for moderately worn cylinder bores. The trade-off is a 2 or 3 percent increase in fuel consumption. Other wearing parts, such as seals, oil pump, and timing drive, should also be renewed. The engine is normally left in place, and machining operations are limited to the cylinder head. Assuming that an overhaul was appropriate in the first place, the effort should be good for 50,000 to 70,000 miles.

A *rebuild* is a much more extensive procedure. In addition to the valve work that is the heart of an overhaul, a rebuild involves regrinding the crankshaft and reboring the cylinders and should include rebuilding the conn rods. The block is chemically cleaned and, when appropriate, main bearing saddles are remachined. Oversized pistons and rings, together with appropriately undersized main and conn rod bearings, are fitted. In addition, a rebuild worthy of the name, includes new:

- ☐ Camshaft bearings
- ☐ Cam drive parts
- ☐ Freeze plugs, and
- ☐ Oil pump

All other wearing parts are inspected, refurbished, or replaced as needed. In theory, a rebuilt engine should be the moral equivalent of a new unit. Practice may fall short of theory, but one can expect 100,000 miles or so of reliable transportation from a carefully rebuilt engine.

VACUUM GAUGE

A vacuum gauge can supplement other test instruments and, in skillful hands, is the practical equivalent of a $6000 electronic engine analyzer. The gauge, which costs less than $10.00, is connected to a source of manifold vacuum. That is, it connects to a vacuum source at any point between the intake valves and the carburetor or fuel injector throttle butterfly. Gauge readings are in inches of mercury (in./Hg). At idle, the throttle butterfly is almost closed and the engine develops high manifold vacuum, usually between 15 and 22 in./Hg. As the throttle plate opens, there is less of an obstruction and the engine breathes easier. Vacuum readings drop toward 0. So long as all cylinders contribute an equal share to the pumping load, the gauge needle holds steady. A weak or misfiring cylinder will cause the needle to dance.

Low Vacuum at Idle: Gauge reading holds steady at between 10 and 15 in./Hg (Fig. 1-15). Possible causes are retarded ignition, late valve timing, weak compression in all cylinders, or air leaks between the throttle plate and combustion chambers.

Erratic Fluctuation at Idle: Needle dances between normal and low readings (Fig. 1-16). Most likely cause is an ignition miss associated with spark plug or spark plug lead to one cylinder.

Regular Drop at Idle: Needle holds steady and drops sharply at regular intervals (Fig. 1-17). Suspect a burnt valve or an improperly adjusted valve that fails to seat.

Drift at Idle: Needle holds steady for a few seconds, drops, and recovers (Fig. 1-18). This condition is known as lean roll and means that the carburetor mixture is too lean to support good combustion. Verify by partially closing the choke in order temporarily to enrich the mixture. This condition can also be caused by a vacuum leak.

Steady Falloff at Idle: Engine seems to "lose ground" and condition worsens as rpm is increased (Fig. 1-19). Look for an obstruction in the exhaust system: the restriction can sometimes be heard, and often develops at the Y-pipe connection on V-8 engines. Retest with catalytic converter disconnected.

Increased Fluctuation as Speed Increases: Rapid, large-amplitude fluctuations mean that a cylinder is out firing (Fig. 1-20). Locate defective cylinder by disconnecting and grounding each spark plug wire in turn. Shorting the ignition on the bad cylinder will have no effect on performance. Check ignition, valves, head gasket.

Fig. 1-15. A steady vacuum gauge reading between 10 and 15 Hg/in. at idle is low, even for "smog motors" with retarded valve timing.

Fig. 1-16. An intermittent fluctuation at idle usually means an erratic ignition miss.

Fig. 1-17. A regular oscillation at idle often translates as a valve problem in one or more cylinders.

Fig. 1-18. A slow drift and recovery at idle is the classic symptom of fuel starvation.

Fig. 1-19. Gradual drop at idle indicates that the exhaust system is restrictive.

Fig. 1-20. Rapid fluctuation at speed means that a cylinder is out.

ABNORMAL IGNITION

Ignition must, of course, be initiated by the spark plug, which is timed to fire just before the piston reaches top dead center. The flame front moves out from the spark plug electrodes and, if things are working right, develops maximum cylinder pressure just as the piston begins to descend on its downstroke. The rate of flame propagation is fairly constant; thus, the ignition is mechanically or electronically advanced as engine rpm increases and less time is available for burning. A distributor with 15 degrees of static, or initial, advance, may have 45 degrees of advance at wide open throttle.

Preignition

Preignition occurs before normal ignition when the air-fuel mixture is ignited by a local hot spot. An overheated exhaust valve; a partially detached, "hangnail" spark plug thread; or even a flake of carbon can be the ignition source. Engine damage may take the form of a holed piston.

Detonation

Detonation occurs after normal ignition. The tag ends of the mixture, compressed and heated by the moving flame front, suddenly explode. The result is often audible: but the sound varies a bit between engines and can be masked by other noises. Vintage engines, with their long, spindly connecting rods, made a pinging noise. The rods vibrated like tuning forks with the force of the blast. Modern engines usually make a deep, staccato, rattling noise.

The effect is to batter the pistons, sometimes knocking a hole in them, more often hammering out the ring grooves and fatiguing the bearings. No engine that detonates will survive long.

Detonation occurs when compression ratio is too high and/or the ignition is too far advanced for the octane of the fuel used. It is aggravated by low turbulence in the intake tract, and tends to occur under low manifold vacuum. Consequently, most engines pass the threshold of detonation when climbing a hill in high gear. Piston speed is low and the throttle plate is open wide. Overheating and high ambient temperatures contribute to the problem.

2
CHAPTER

Logistics

THE MECHANICAL ASPECTS OF ENGINE REBUILDING ARE RELATIVELY STRAIGHT-forward and require no more than ordinary skills. The real problems are logistical: technical data, tools, replacement parts, and machine work must be combined into a coherent whole. For convenience, the project can be divided into eight stages:

- ☐ Gathering information
- ☐ Initial cleanup
- ☐ Preliminary disassembly
- ☐ Engine extraction
- ☐ Disassembly
- ☐ Replacement parts and machine work
- ☐ Assembly
- ☐ Installation
- ☐ Start-up and tuning

GATHERING INFORMATION

The work is only as good as the quality of information that supports it. This book will be useful, I hope, as an introduction to the rebuilding process. But no general treatment of the subject can take the place of a factory manual for the engine at hand.

There are real factory manuals and compendiums of factory manuals. The real thing is better, since it covers the product in greater depth of detail and is a more direct reflection of factory thinking. Current and very late model manuals can be purchased from auto dealers, and some companies, such as Ford

Motors, can supply books for older models. An array of training materials can be ordered from Ford Service Publications, P.O. Box 07150, Detroit, MI 46207.

Unfortunately, there is little direct-factory support for vehicles old enough to need rebuilding. One must search libraries and used-book stores for these out-of-date, but vital, manuals. Occasionally the search turns up a treasure, such as original parts books and special factory publications on emission control, fuel injection, and other complex systems.

Compendium books, published by houses such as TAB, Chilton, Motor, and Clymer's, may be all that is available. Perhaps the most impressive of these is the series published by Haynes in Great Britain. The editors depend heavily upon factory sources (recasting the material into good English) and supplement this information by dismantling examples of the vehicles written about. HP publishes a series of books on rebuilding and modifying Ford, Chevrolet, Nissan, and other popular engines. These books combine factory-supplied information with trade practices developed by working mechanics.

Further information can be gleaned from specialty and trade periodicals. Most big city libraries subscribe to *Hotrod, Car Craft*, and similar hobbyists' publications. The emphasis is upon performance, but the techniques described apply to all engines. Trade journals, such as *Motor's Magazine*, are more to the point and include abstracts of factory tech bulletins. Back issues of *Popular Science* and other how-to magazines are also worth consulting.

INITIAL CLEANUP

Once you have collected the requisite information, the next step is to clean the assembled engine. Much grime will remain, shielded behind the manifolds and other parts, but the effort is worthwhile.

Assembled parts can be cleaned with any of the various aerosol products sold for this purpose. Gunk Concentrate, mixed in a 1:5 ratio with kerosene or Varsol and applied with a stiff-bristle brush, is as effective as the spray-type cleaners and considerably less expensive. Gunk Concentrate can be ordered through car parts stores for about $3 a pint. All such engine cleaners' products work by converting soft grease to soap, which must then be hosed off. Road grime of the kind not associated with grease responds to a detergent and hot water. Tide and Sears brand detergents seem to work best.

Water-flushed solvents do a good job on the sludge that accumulates on the undersides of valve covers, timing chain cases, and oil pans, and can be used on engine internals. However, parts so treated must be thoroughly dried and oiled. Clean parts rust within minutes. Final cleanup requires liberal amounts of kerosene or Varsol. Some mechanics wipe bearing bores and critical gasket surfaces with lacquer thinner just prior to assembly. You will also need a supply of clean shop rags and lint-free paper towels, of the kind sold at auto parts houses.

An air compressor is useful for cleaning oil passages, but is not absolutely vital.

INITIAL DISASSEMBLY

There is no general rule covering which parts should be removed from the engine before it is lifted. On a Honda Civic, one begins with the right headlamp rim. A domestic pickup truck engine can be extracted in one bite, merely by disconnecting the plumbing, wiring looms, exhaust manifolds, motor mounts, and transmission.

To start with, you will need a good, but by no means professionally complete, collection of basic tools. Quality does not have to be the best—upper middle range tools, available from auto parts stores, are entirely adequate. Many amateur and professional mechanics favor Sears Craftsman tools, which are readily available and supported by a no-questions-asked warranty.

A ½-in. drive socket set, which should include a cachet handle, breaker bar, and assorted extension bars, is the workhorse of engine service. Foreign engines require metric sockets; late model domestic engines demand both metric and SAE sizes.

A ⅜-in. drive socket set is a virtual necessity if you are working in crowded engine compartments. The set should include a short (1½-in.) knuckle-saving extension bar, a "wobble joint," and two adapters (⅜-in. female to ½-in. male and ½-in. female to ⅜-in. male) so that drivers and sockets can be interchanged.

End wrenches, including duplicates in popular sizes (7/16, ½, and 9/16 in. or 12, 13, and 14 mm), together with flat and Phillips screwdrivers, will find plenty of use. Vise-Grips—the trade name has become generic for all locking pliers, regardless of manufacturer—are backup tools, used when all else fails. An adjustable wrench, large enough to have authority, is also useful.

Major threads—head bolt, main-bearing-cap and manifold-mounting threads—need to be chased with a bottoming tap, distinguished from the ordinary variety by a blunted end and constant diameter. A good machine shop will attend to this. Otherwise, you will have to purchase the necessary taps, which are not easy to find in metric sizes. In a pinch, you can convert a plug tap to a reasonable facsimile of a bottoming tap by grinding the pointed end off.

Every engine rebuild presents unanticipated difficulties that can only be solved by additional tool purchases. For example, you may encounter special fasteners such as Reed & Prince, clutch-head, or unusually large recessed-head bolts, often in metric sizes. Budget accordingly.

ENGINE EXTRACTION

Lifting an engine is inherently dangerous and made more dangerous by the need to work from the underside of a vehicle raised off its wheels.

The safest and most convenient way to lift an engine is to use what is variously described as a portable jib crane, a shop hoist, or more commonly, a cherry picker (Fig. 2-1). The jib is raised and lowered by a hydraulic ram, and the whole assembly pivots on casters. Two-ton capacity cherry pickers, mounted on trailer axles, can be rented for about $12 per hour. With some planning, you should be able to keep rental time down to no more than 3 hours.

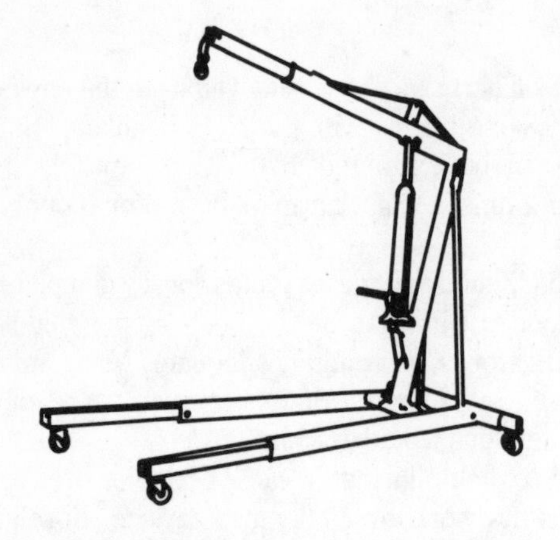

Fig. 2-1. A mobile shop crane is more than worth the rental and must be used on some vans.

An alternative is to construct a stationary A-frame, preferably from steel, and provide enough vertical reach for the oil pan to clear the fenders. A chain hoist or less expensive cable hoist (Fig. 2-2) provides the lift. (Northern Hydraulics, 1-800-533-5545, advertises a 2-ton cable hoist for $10.95.) Once the engine is extracted, the car is pushed out from under the frame and the engine lowered to a dolly or a stand.

Of course, there are other ways to raise an engine. But however you go about it, consider that automobile engines are heavier than they appear. As a base number, 1 cubic foot of cast iron weighs about 475 lb. Early V-8 engines, such as the 1949 Cadillac ohv (overhead valve), had an all-up dry weight of more than 700 lb. Their descendants are lighter, but some must be extracted with the transmission attached, and any engine can "hang up" during removal, transferring vehicle weight to the lifting apparatus and its supports.

A garage rafter will not do.

You will also need at least one 5- to 6-ft. length of chain or, better, an engine sling, available from Northern Hydraulics in a chain version or, as a cable sling, from Snap-On Tools (Fig. 2-3). Some mechanics prefer a beam-type sling, as illustrated in the following chapter (Fig. 3-9).

A garage-type floor jack is a practical necessity. Every mechanic, no matter how occasional, should have access to a 1½- or 2-ton model (Fig. 2-4). The best jacks are still made in the USA, and while the initial cost is more than

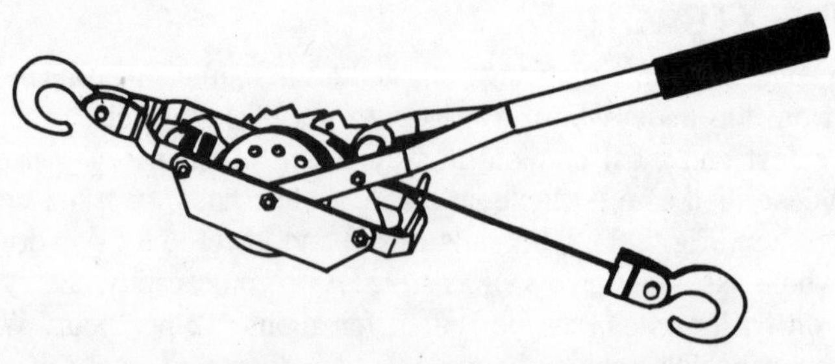

Fig. 2-2. Nonprofessional mechanics generally lift the engine with a cable hoist, which can be purchased for less than $15.00.

imports, American jacks are repairable. Leave the bumper jack in the trunk where it belongs.

A small hydraulic or screw-type jack is a useful adjunct to the floor jack. Adapters are available to convert floor jacks to transmission jacks, but the sample I purchased leaves much to be desired. If the transmission is heavy and there is some danger of getting hurt, rent a proper transmission jack.

Any jack can slip or fail; hydraulic jacks are subject to total collapse in the event of seal rupture. The weight of the vehicle must be borne by jack stands, footed solidly on a hard surface and bearing against strong points on the underside of the chassis (Fig. 2-5).

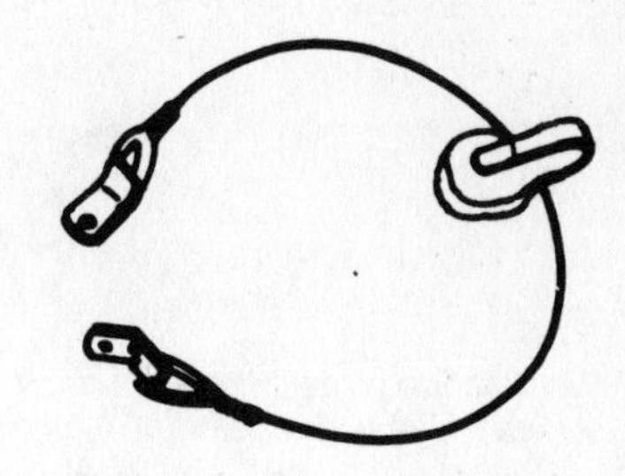

Fig. 2-3. A sling can simplify extraction in those cases where the engine must remain reasonably level. Nose-up extractions, as when the transmission (rear-wheel drive) comes out with the engine, are best done with a bar-type sling or two chains.

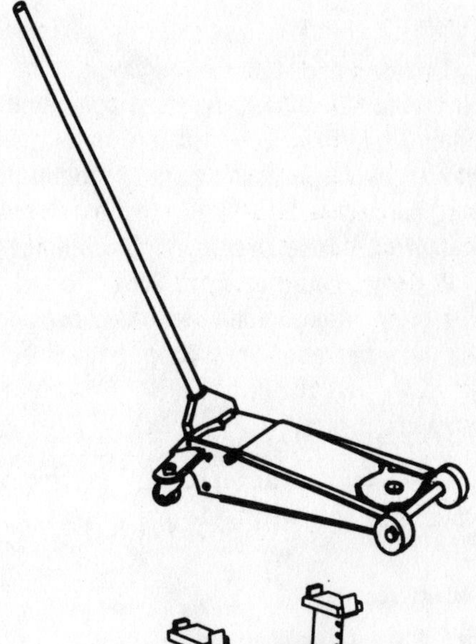

Fig. 2-4. A hydraulic service jack is one of those tools that everyone who works on cars should have. Purchase a standard-brand 1½- or 2-ton model.

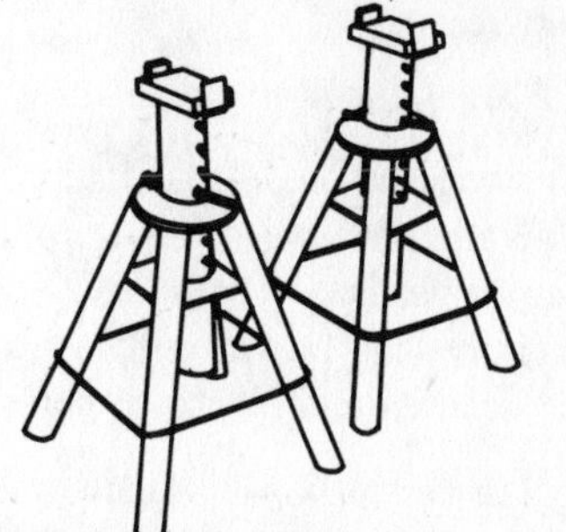

Fig. 2-5. Purchase or borrow a matched set of adjustable jack stands, with a rated capacity comfortably larger than required by vehicle weight.

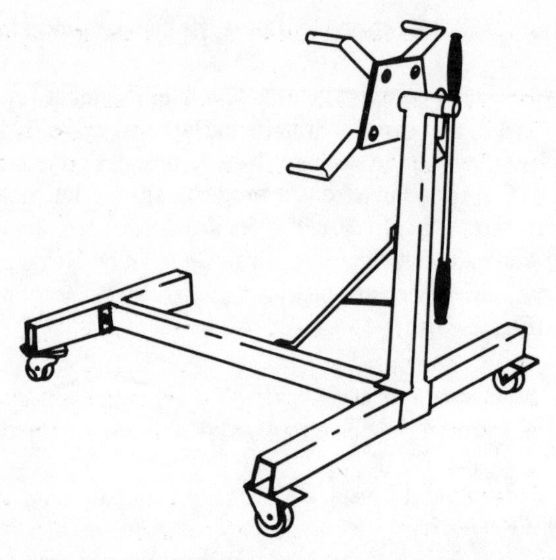

Fig. 2-6. If one is to use an engine stand—and it is a marked convenience—it pays to purchase a good one, adaptable to a wide range of engines, with a worm gear cranking mechanism and a rated capacity of at least 750 lb. Also note that geometry of engine stands makes them liable to tip over.

Engine stands fall into the category of nice, but not necessary. Cost ranges from about $50.00 for the most basic unit to more than $300 for professional models (Fig. 2-6). If you plan to build more than one engine, you might think seriously about purchasing one, recognizing that the cheaper units may be hard put to support a big V-8 and that some cutting and fitting is usually necessary to adapt the stand to less popular engines.

A dolly, bolted up from 2×4s and mounted on casters does not take the place of an engine stand, but does provide a relatively painless way to move heavy engines about the shop.

DISASSEMBLY

Published sources cannot be trusted to convey all the information you may need. A mechanic soon learns to make notes.

Taking Notes

Critical areas for notes include:

- ☐ Accessory drives and support brackets
- ☐ Variations in bolt length, often encountered on water pumps and cylinder heads
- ☐ Locations of wiring hangers and other small parts
- ☐ Vacuum and electrical hookups and routing

Most of these relationships can be sketched; vacuum and electrical connections should be tagged as described in the next chapter and shown in the sketched layout. Some amateur mechanics make a photo record of each stage of disassembly.

The immediate reason for taking notes is to be able to get the engine back together again with a minimum of dither; the larger purpose is to come to an understanding of the logic of the engine. The fuller your understanding, the fewer notes you need to take.

Parts Storage

Parts should be cleaned with kerosene or Varsol and stored in some semblance of their original order. The order does not have to be perfect (as in an exploded view), but related clusters of parts should be stored together. Allocate enough space for storage so that you can retrieve individual parts without jumbling the others.

Heavy castings take care of themselves; smaller parts should be stored in cardboard cartons, together with their fasteners. Food cans make excellent containers for bolts, nuts, washers, and other small parts, especially if the cans are sealed with plastic lids. Baggies can also be used and have the added advantage of transparency. The baggies can be labeled—"pan bolts," "carburetor hardware," etc.—with a Sharpie pen. Bolts can be temporarily reinstalled on castings that will not be sent out for machine work.

Special Tools

Purchase or rent special tools as the need arises. For example, worn cylinder bores often leave a ridge above the area of ring travel, which you need to dress with a reamer before extracting the pistons. It is possible to pry some harmonic balancers free of the crankshaft hub; others have to be pressed into place and require a hub-type puller, which rents for about $3 a day.

INSPECTION

Inspect each part for usability, including exhaust manifolds and other castings. We are looking for cracks, surface irregularities, and evidences of wear. A jeweler's eye loupe (available from photographer's supply houses for about $5) is invaluable for this purpose.

The machinist will double-check bearing clearances, but it is useful to be able to make your own measurements. The most cost-effective measuring instrument is a 6-in. dial caliper, which, for decent quality, costs about $60 new or half of that used (Fig. 2-7). Vernier calipers are less expensive, although more difficult to read. Caliper accuracy does not match that of a properly calibrated micrometer, but suffices for most jobs. You will also need a set of feeler gauges in metric or inch denominations. Rigged with a suitable fixture, a feeler gauge can serve as a dial indicator to check flywheel and other rotating-part runouts.

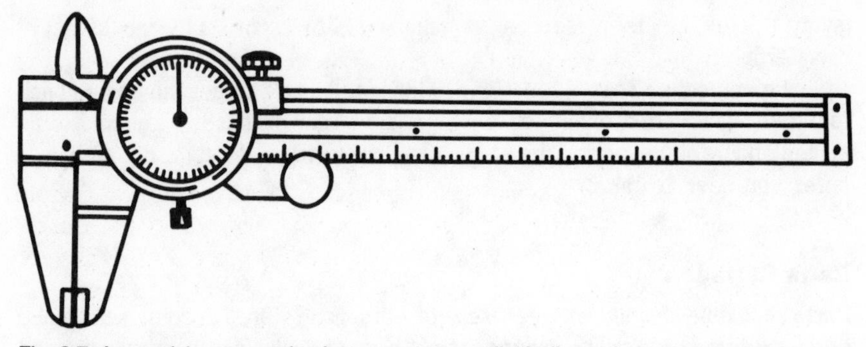

Fig. 2-7. As precision measuring instruments go, a dial indicator is a rough and ready tool, sufficiently accurate to raise questions that can be settled with a micrometer.

PURCHASING PARTS AND MACHINE WORK

At this point we can make a list of parts that will be needed and begin to get an accurate idea of the cost of the project.

Replacement Parts

Types. New replacement parts fall into two overlapping categories: OEM, or original equipment manufacturer, and aftermarket. Until fairly recently, OEM parts were available only from auto dealers and were almost uniformly superior to aftermarket parts. Neither of these distinctions holds true today. Some OEM parts are available in parts stores as auto manufacturers reach for a larger market outside the dealer network and as more parts are "outsourced." TRW, Bendix, Felpro, and other major aftermarket manufacturers also supply the original equipment market. Some aftermarket pieces are superior to OEM; some are much worse.

Sources. Auto dealerships are indisputably the best sources of parts for vehicles they sell. However, prices tend to be high and most mechanics view the dealer as a last resort. Discount houses generally have the best prices. However, selection is skewed toward fast-moving items. Emphasis is on the so-called A inventory—meaning vehicles built since 1978. Parts quality ranges from very good to poor.

Neighborhood parts stores, sometimes organized into local chains, occupy the middle ground between dealer resources and discount house bargains. Selection is wide and countermen are specialists in ferreting out elusive parts. Some of these outlets rent specialized tools.

Automotive machinists also supply parts, and many expect to make a parts sale on every job. Sometimes their prices are competitive and sometimes not. If you buy from another source, bring the replacement bearings and pistons to the machinist for fitting. He should have the opportunity to validate running clearances for the parts that will be used.

Do not overlook the local wrecking yard. It is the best source of castings and other heavy parts that are not available in the aftermarket and that would be prohibitive if purchased from a dealer. Informal, somewhat disorganized operations ("Well, we probably got one of those back there—take what you need!") usually have the best price, which in no event should be more than half the cost of a new part.

Buying. As noted, quality varies in the aftermarket, and one should always specify name brand parts, even though you might pay a little more for Perfect Circle piston rings or Michigan bearings. Some of the worst-quality aftermarket pieces do not carry a brand name, in the sense that the brand is the signature of the marker. The name on the box is merely the name of the importer.

The basic parts identification consists of vehicle make, model, and engine displacement (e.g., Chrysler Aries, 1983, 135 cubic in.). This may be supplemented by build date (which can be decoded with a suitable factory manual) and by application (usually California or 49-state). In recent years there has been a tendency among manufacturers to make unannounced engineering changes which can only be detected by comparing the new parts with the old. Some differences are superficial and may represent design improvements; dimensional differences are usually critical.

Dealer-purchased OEM parts are usually okay and appropriate for the application. However, it should be mentioned that counterfeit parts are in circulation. These parts are disguised as OEM and packed in factory look-alike boxes. Some are really good forgeries that nothing short of metallurgical analysis can detect. Others are quite sloppy in terms of finish and packaging. It is safe to say that none of them works as well as the genuine item.

Choosing a Machinist

The quality of machine work varies with the individual who performs it and with the contractual arrangements made to pay for it.

The low-bucks, high-risk approach is to farm out the work on a per-job basis. Most neighborhood parts stores are associated with a machinist, who may be on premises. Some factory rebuilders will refurbish loose parts. In either case, the cost is set by formula, typically about $80 for reworking a set of V-8 heads and $4 a hole for reboring cylinders. Responsibility does not, of course, extend beyond the work done.

A better, but considerably more costly, approach is to turn the work over to a premium "repair-and-return" shop. You bring in the cylinder head (with valves assembled), the block, piston assemblies, and crankshaft, together with any other engine parts that appear to need attention. The shop takes responsibility for the engine as a complete entity, performing all the machine work, selecting the replacement parts (unless you specify otherwise), and usually extending technical advice during assembly.

Premium shops can be found in any large city and wherever there is an industrial base that justifies the cost of professional services. Look for a shop that has been in business a long time, that has heavy-duty engine blocks lying around, and a friendly and attentive counterman.

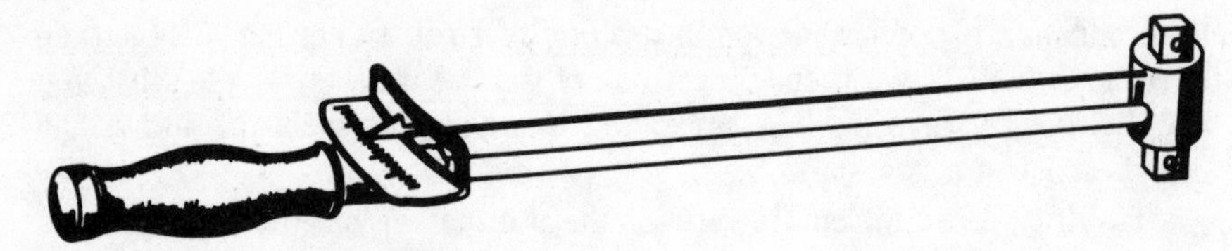

Fig. 2-8. Deflecting-beam torque wrench.

ASSEMBLY

Assembly is the nexus of the whole process, where refurbished parts, new parts, and technical knowledge come together.

Special Tools

In addition to the usual hand tools, the measurement instruments already mentioned (never trust anybody), and a pumper-type oil can, you will need at least two special-purpose tools—a torque wrench (½-in. drive, 150-ft.-lb. capacity) and a piston ring compressor. You will probably need a ring expander, although some custom machine shops deliver the pistons with rings mounted. An inexpensive (about $25.00) deflecting-beam-type torque wrench, such as the Sears Craftsman 9 HT 44642, does an acceptable job (Fig. 2-8). The wrench has both ft.-lb. and Newton-meter scales. More sophisticated click-stop, dial-reading and electronic wrenches are easier to use, but must be periodically recalibrated against a known-good standard. *Recalibration* of beam-type wrenches merely means that the pointer is bent to register zero under no load. Band-type piston ring compressors retail for less than $5.00, as do ring expanders.

Other tools can be purchased or rented as the need arises. Used pistons must have their grooves cleaned prior to installation: a thorough machinist will see to this detail; otherwise you must clean the grooves with a broken piston ring or rent a groove scraper.

Other special tools fall into the category of luxuries. Oil seal installation tools are a case in point. A piece of masking tape will protect the seal lips from the sharp edges of keyways and a suitably dimensioned wood plug or PCV pipe can serve as a driver.

Absolutely vital special tools, such as the trick belt tensioners used on some overhead cam engines, can usually be purchased from the aftermarket. Failing that, one must hire that portion of the work out to a dealer mechanic.

Supplies

RTV (room temperature vulcanizing) silicone is often specified as a "bridge" between the ends of conventional gaskets. Applications include Chrysler small-block V-8 manifold end seals, pan gaskets on most Pontiac V-8s and as front and rear seals on 1980–81 Cadillac plastic oil pans. This product has dozens of other uses, but must not be exposed to gasoline or exhaust temperatures. Nor should RTV be applied on valve cover gaskets, where it acts as a lubricant.

Another useful product is liquid gasket sealer (''shellac''), which is brushed or sprayed on die-cut gaskets for additional protection from leaks. Gasket sealer is incompatible with some gaskets, and use should be limited to those parts that are certain to stay together. Most common applications are water-pump and thermostat-cover gaskets.

Some fasteners extend into the water jacket and must be sealed to prevent leakage around the threads. Fel-Pro Plia-Seal or General Motors PN 1050026 work fine; old-timers swear by Permatex No. 2.

Most mechanics continue to rely on motor oil as assembly lubricant, but there is an increased tendency to use an assembly grease, such as Lubriplate 105. This or a similar lubricant is indispensable for installing a new camshaft. Michigan Bearing Guard is specially formulated to prevent score on start-up.

The hot-dip process that removes carbon and scale from iron parts also takes the paint off. Some machine shops repaint these parts; others do not, reasoning that the customer should be able to see what he is buying. Any good quality paint suffices; original-color engine enamels can be purchased as aerosols from car parts houses; paints for many vintage engines are available from Bill Hirsch (396 Littleton Ave., Newark, NJ 07103). Hirsch also sells a black gloss exhaust manifold paint, which he claims is second only to porcelainizing. The manifold must be sandblasted before application.

INSTALLATION

Engine installation is the reverse of extraction, with the proviso that the engine should be assembled as completely as possible. Ideally, the engine would be mounted on a test stand and tuned before it was put back into the vehicle, but this is out of the question for most of us. The final adjustments are made with the engine installed in the vehicle.

3
CHAPTER

Engine Removal

THIS CHAPTER DESCRIBES A GENERALIZED PROCEDURE FOR REMOVING ENGINES from front-engine, rear-wheel-drive passenger cars and light trucks. While some of this material applies to front-drive vehicles, no general approach works for all makes and models, and you will need to consult the appropriate factory manual.

The author hopes that readers will not be put off by the number of cautionary notes in this chapter. Extracting (and installing) engines is a fairly hazardous activity, and all of us have had our near misses. Stop the proceedings if you feel uncomfortable, stay out from under overhead loads, and wear safety glasses. Mechanics are not praised for their bravery.

OVERVIEW

Lift the engine as a complete assembly, minus only those components that physically impede extraction or that might be damaged during the process. The power steering pump and ac (air conditioning) compressor stay in the vehicle with their plumbing intact. Vulnerable parts, such as fan, distributor cap, alternator, and radiator, should be removed. Space considerations may dictate removal of other parts, such as the evaporative emission system on some Chrysler products and Camero/Firebird valve covers.

Full-sized pickup trucks and other vehicles with generous hood space give the option of extracting the engine separately or with the transmission attached. It is generally easier for the do-it-yourself mechanic to leave the transmission in place, provided the bellhousing bolts are accessible.

The usual procedure is to separate the components at the engine/bell housing interface. The engine is then nudged forward to disengage the pilot shaft, and then lifted out with crankshaft parallel, or nearly parallel, to the ground. Lack of clearance in front of the engine sometimes makes it necessary to disengage the clutch shaft by dropping the transmission. Only a little more work is involved, and the transmission can then be used as an awkward, but definitive, tool to establish clutch alignment.

When the transmission and engine must come out together, tilting the assembly at a 45-degree angle will clear the front-end sheet metal of many full-sized Ford cars, some GM products, and a few late-model Chryslers.

PRELIMINARIES

Drain the coolant and oil, preferably at some distance from where you will be working. Empty the cooling system both at the radiator (if a drain cock is not present, remove the lower radiator hose) and at the engine. Tighten pan and block drains now, while you're thinking about it. A fingertight drain plug is easy to overlook and can lunch the newly rebuilt engine.

Using a Magic Marker or scribe, outline the footprint of the hood hinges. This will simplify hood alignment during assembly. Remove the hinge bolts and, with the aid of a helper, carefully lift the hood off the vehicle, keeping the hood clear of the windshield. Hinge bolts, which are special body fasteners, can be run back into the hood for safekeeping. If storage space is tight, you can park the hood on top of the vehicle.

TOPSIDE

Remove the battery, noting battery cable polarity. With marker pen and masking tape in hand, begin the work of disconnecting those vacuum, electrical, fuel, and water lines that connect the engine and chassis (Fig. 3-1). Work deliberately,

Fig. 3-1. Label electrical, vacuum, and fuel lines.

from the fire wall forward on one side of the engine bay and back around to the other side. Tape and number each hose, line, and wire in the order of encounter. Hose #1 goes to connection #1, wire #2 to terminal #2, and so on. Later, during assembly, one merely follows the sequence of numbers and knows immediately when a connection has been skipped. If things begin to blur, pause and make a detailed sketch, referencing some prominent engine feature. Photographs are also helpful.

CAUTION—Fuel lines for Ford, GM, and Bosch-pattern electronic fuel injection systems hold constant pressure (ranging from 15 to 40 psi). Ford thoughtfully includes a bleeder plug at the engine-side filter on some models. Special tools are recommended for other EFI systems. Although the practice is dangerous and not recommended, connections are sometimes broken "wet" on cold and battery-less engines. Protect your eyes and contain the spill with a shop towel wrapped around the line. Plug engine and tank lines to prevent contamination and siphoning.

Hardware

Remove the air cleaner, labeling the attached vacuum lines and their connection points. Disconnect the throttle cable or rod at the carburetor, together with the associated bracketry (Fig. 3-2). The linkage is secured to the throttle by a lock clip (a wrecking yard can replace a lost clip). Disconnect cruise-control cable or chain at the throttle and the kickdown rod that runs between the throttle and automatic transmission.

Remove the radiator hoses. Special pliers are available to defang spring clamps, although a pair of Vise Grips works almost as well. Slip the clamp several inches behind the connection and twist the hose to break its seal. Hoses that have vulcanized themselves in place may be persuaded with a pair of water-pump pliers. Do the same for the heater and automatic transmission coolant hoses. Loss of transmission oil through siphoning can be avoided by disconnecting the hoses at the transmission (necessary if the transmission is to come out with the engine) or by connecting the free ends of the hoses with a length of 5⁄16-in. diameter tubing.

Remove the fan shroud and radiator, typically secured in a rubber saddle by clamps (crossflow type) or bolted solidly to the forward bulkhead (downflow). Send the radiator out for professional cleaning.

Unbolt the fan assembly from the water pump. Steel and fiberglass fans are subject to fatigue cracks at the blade roots and mounting bolt holes. Discard any fan with this damage. Viscous-drive fans should be stored vertically in the "as-car" position to avoid leaks (Fig. 3-3).

Most engine can be extracted with the exhaust manifold(s) in place. Disconnect the manifold-to-exhaust-pipe connection, working, as convenience dictates, from the engine bay or from the underside of the vehicle.

CAUTION—Wear safety glasses when loosening exhaust-pipe connections from below.

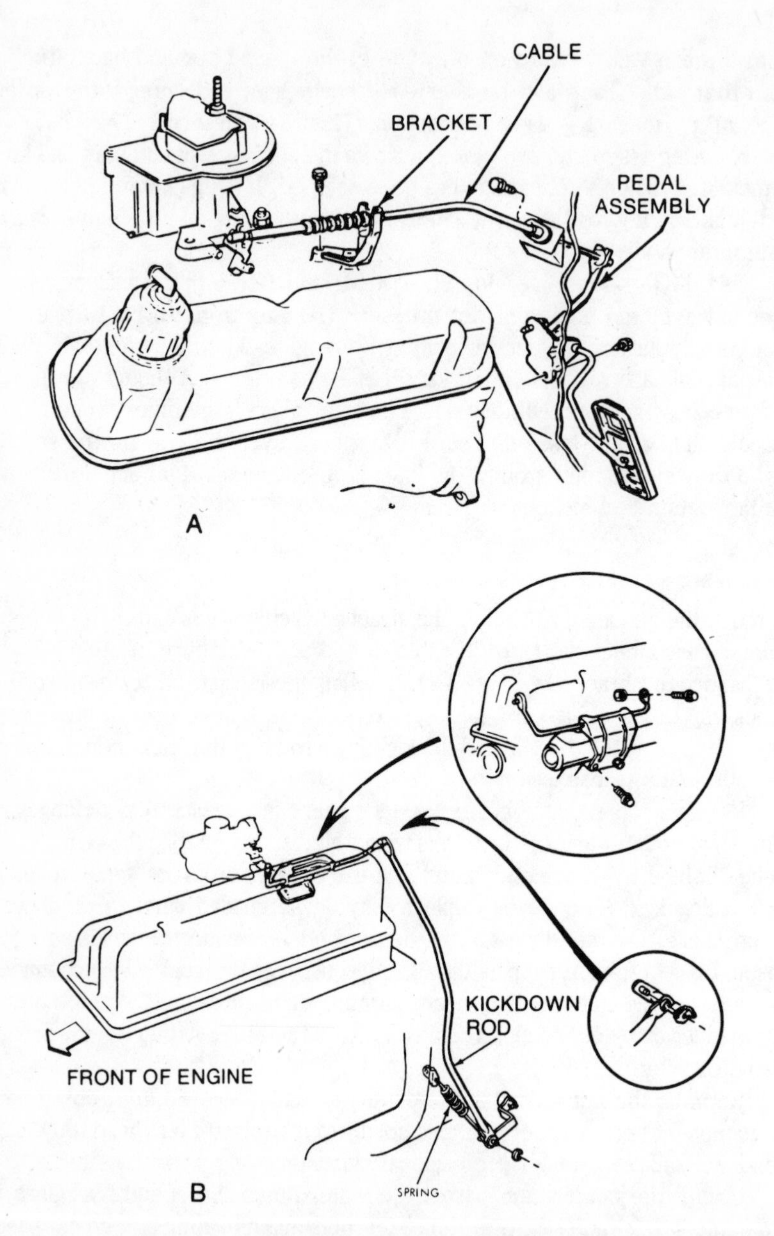

Fig. 3-2. Most modern engines use a flexible cable for throttle control (A) and a rod to signal automatic-transmission downshifts (B). (Courtesy Ford Motor Co.)

The crossover pipe, which loops under some V-8 blocks, should be removed. Exhaust manifold bolt breakage can be minimized by:

- ☐ Soaking the bolts with penetrating oil for several hours (or days) before disassembly,
- ☐ Shocking the bolt heads with a hammer and heavy punch, as if you were driving a nail,

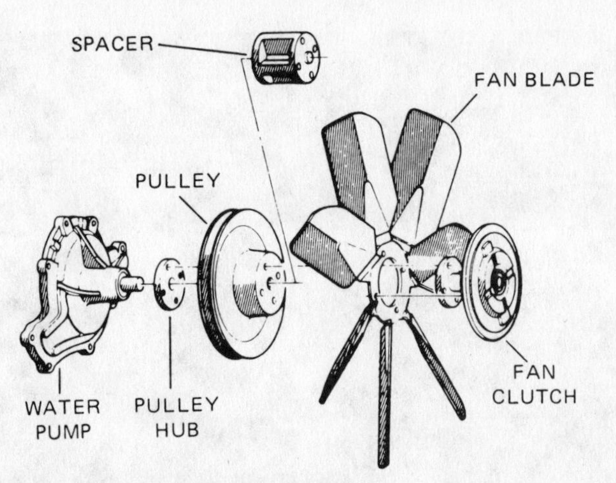

Fig. 3-3. Viscous-drive fan clutch should be stored in the as-car position to avoid loss of fluid. (Courtesy Sealed Power Corp.)

- ☐ Tightening the fasteners a tiny fraction of a turn to break the rust bond and alternately tightening and backing off as the fasteners are undone. Use a six-point socket and, if the fastener has lost dimension, force a smaller wrench size over it. A slightly smaller metric socket can sometimes be substituted for nominally correct American size and vice versa.

Slack of the drive belts, remove them, and detach the air-conditioner compressor from its mounting bracket.

CAUTION—Do not disconnect Freon lines going to the compressor or other air-conditioning components. All lines are pressurized and liquid Freon is extremely hazardous, causing burns and even permanent blindness. Carefully, without forcing the hoses, reposition the compressor up against the inner fender, and lash it in place.

Detach the power steering pump at the engine—again without disconnecting the hoses—and secure the unit to the chassis or fender splash panel. Other belt-driven accessories may be removed at this time.

If the hoses feel spongy, make a note to replace them when the engine is up and running again. Disconnect the return hose and crank the steering wheel to engage the pump. After installing the new hose and an optional filter, saw the wheels back and forth to purge air from the system.

Miscellaneous

Remove the automatic transmission filler (dipstick) tube, usually secured by a bolt at the top and a plug-in connection below. The area between the top of the bellhousing and the floorpan must be free of clutter to allow the engine/transmission to be raised.

Fig. 3-4. Never play cards with a man named Doc, eat at a cafe called Mom's, or trust a jack.

DOWN UNDER

Block the rear wheels, and using a floor jack, raise the front end of vehicle high enough to provide working room under the transmission. Without getting under the car, position jackstands under chassis strong points or major suspension members (Fig. 3-4).

CAUTION—Do not work under an unsupported jack.

Slowly lower the jack, repositioning the stands as necessary to keep the load path vertical. Test the security of the arrangement by attempting to rock the vehicle from side to side and fore and aft. Do not crawl under a vehicle that you can shake, especially if you will be under there applying torque to fasteners.

Make a reconnaissance to determine what must be dismantled. The transmission will separate either at the bellhousing or at the bellhousing/engine interface, as determined by bolt access. You can reach upper bellhousing bolts from the engine bay or from under the vehicle by snaking a long extension bar and socket over the top of the gearbox.

Light-duty manual boxes can be manhandled out; automatics and heavier standard transmissions must be supported by a jack. Most amateurs use a floor jack and hope that the transmission does not fall off. A better and safer approach is to rent a proper transmission jack.

DRIVESHAFT

Mark the rear U-joint as shown in Fig. 3-5 as an assembly guide. Undo the four bolts—noting that they are special, high-strength fasteners—remove the center bearing bracket, and pull the shaft out of engagement with the transmission.

Fig. 3-5. All rotating elements—flywheel, torque converter, and rear U-joint (shown here)—should be marked for proper assembly.

TRANSMISSION

Manual boxes can be drained, since it is probably time for an oil change anyway. Securely replace the drain plug and tag the transmission, indicating that it is dry. To detach an automatic transmission from its mounts, you can partially drain it by placing a catch pan under the fluid cover and removing the cover bolts. Work from the rear bolts forward. Secure the pan with two bolts and make note to replace the pan gasket and filter. Drain the torque converter by removing the access plate and rotating the engine until the drain plug is down.

CAUTION—Turn the crankshaft in the direction of normal rotation on engines with belt-driven camshafts. Reversing rotation can disengage the belt.

You can manipulate the transmission without first draining the oil, by disassembling the front U-joint and using the yoke as a stopper (Fig. 3-6).

The transmission crossmember, the pressed-steel beam upon which the transmission rides, usually bolts to the frame. The rear of the transmission will drop slightly, pivoting on the front engine mounts, when this crossmember is unbolted. Note any shims that may be present between the crossmember and the transmission mount, and examine the rubber mount for separation at the vulcanized joint, cracks, or other evidence of failure. Miscellaneous items, such as clutch linkage, shifter rods, speedometer cable, and backup switch leads must also be disconnected. At this point, the transmission can be extracted in-unit with the engine.

If you intend to drop the transmission as a separate entity, disconnect the starter wiring, making a sketch of the color-coded connections. GM vertical-bolt starters employ shims to establish the depth of tooth contact between the starter pinion gear and the flywheel. Note how many 0.015-in. shims are pres-

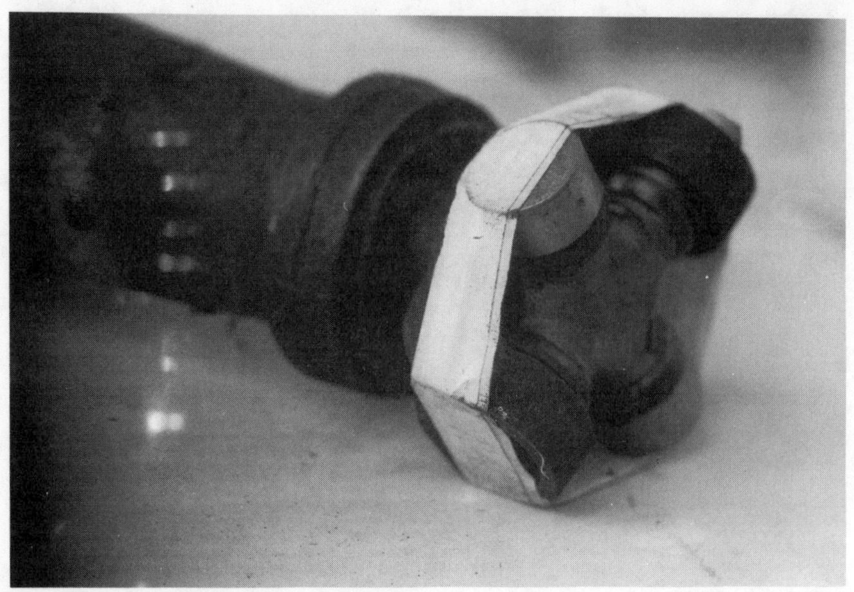

Fig. 3-6. U-Joint yoke can prevent oil spills.

ent. A missing shim will allow the pinion to move deeper into engagement with the ring gear teeth. The pinion may give a little whoop of protest after the engine starts and before the pinion retracts. A superfluous shim will move the pinion away from the flywheel and produce a high-pitched whine during cranking.

Remove the inspection plate on automatic boxes and, using an open-end wrench, unscrew the bolts securing the torque converter to the flex plate. A long-reach, thin-profile wrench of the type used to adjust tappets may be required. Rotate the crankshaft (remembering the caution above) for access to the uppermost bolts. Store these bolts in a labeled container. Use a holding fixture, such as the one shown in Fig. 3-7, to secure the torque converter to the input shaft.

Support the transmission on a jack and, working from the top down, remove the bellhousing-to-engine or transmission-to-bellhousing bolts.

Note: The weight of the transmission must be borne by a jack; in no case should the transmission be allowed to cantilever off the back of the engine, supported only by the input shaft. This is dangerous—the transmission may fall—and is guaranteed to distort the clutch pressure plate and do violence to the input-shaft or transmission-oil-pump bearing.

Upper bellhousing bolts can usually be reached from the engine bay; otherwise, extract the bolts from below, snaking a long extension bar and socket over the top of the transmission.

Tug and wiggle, coupled with small adjustments to jack height will usually disengage the alignment pins. If that fails, double-check that all fasteners are out and gently pry the castings apart. A second jack, placed under the rear of the oil pan, may help stabilize the engine and take some downward thrust off

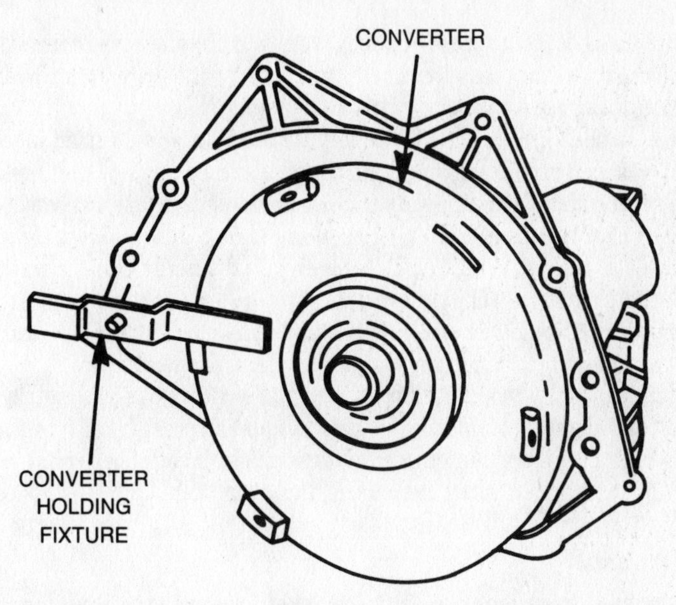

Fig. 3-7. An easily fabricated holding fixture will prevent the torque converter from falling off the input shaft.

the transmission input shaft. Roll the transmission back and out from under the car.

TOPSIDE AGAIN

Lower the vehicle off the jack stands. At this point, all that secures the engine are the front motor mounts. A jack may be used under the engine to limit torsion on the motor mounts.

Attach the hoist to the engine with a short length of chain, sling, or spreader bar (Fig. 3-8). A spreader bar absorbs side forces that would otherwise be applied to the hold-down bolts. Commercially made lifting tackle should incorporate a comfortable safety factor: however, it is worth considering that wide spreads—

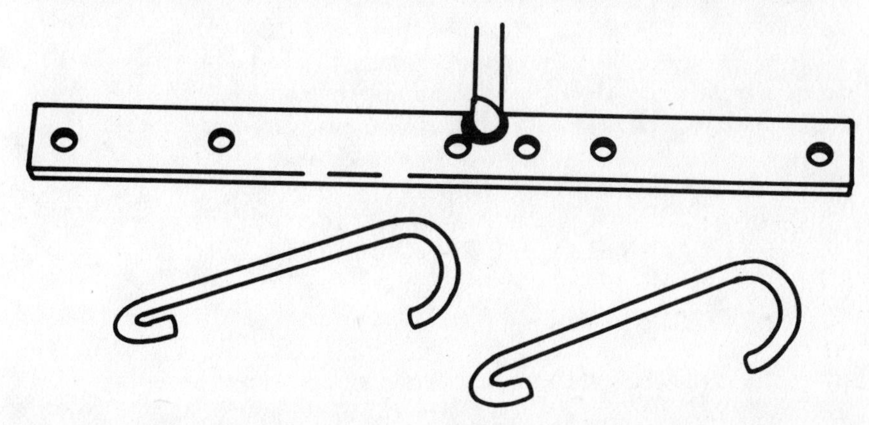
Fig. 3-8. A spreader bar can be used with short lengths of chain or hooks.

angles in excess of 30 degrees—materially reduce the strength of the associated chain or wire rope. Protect wire rope from contact with sharp edges by inserting wood blocks between the wire rope and casting.

Some engines are equipped with lifting hooks; others demand ingenuity to find two balanced strong points. Bolts on opposite corners of the intake manifold can serve as attachment points for V-8 engines; fore-and-aft head bolts are usually the best bet for in-line engines. Bolts should be run down tight against the chain or at least to a depth of 1.5 times bolt diameter. Heavy washers will prevent the links from slipping over the bolt heads.

Tension the rigging to see that it stands clear of vulnerable castings and that it allows enough lift for the oil pan to clear the fenders.

A bare engine, without transmission, usually comes out horizontally or with the front end slightly elevated. The hoist should act on or just forward of the engine center of gravity. An engine with transmission attached must be angled sharply up, with most of the weight on the rear.

Motor Mounts

Once unbolted, most motor mounts will skid forward with the engine during the extraction process. Others terminate in a stud that must be cleared before the engine can move forward and out.

Extraction

Make one last check to determine that all is disconnected, and slowly raise the engine. Move the lifting apparatus or vehicle as necessary to clear the fire wall and forward sheet metal.

4
CHAPTER

Cylinder Head Removal

THE CYLINDER HEAD IS THE FIRST MAJOR COMPONENT TO COME OFF AND, ONCE removed, gives the first real insight into engine condition. Now we can examine the combustion chambers and make some determination of cylinder bore wear.

What follows is a general disassembly sequence, that applies to most engines, most of the time (Fig. 4-1). Modify the procedure as necessary. For example, the Ford 390 family uses a ''captive'' distributor that must be removed before the intake manifold can come off. Keep parts and fasteners together, as described in Chapter 2, and pause to make notes when things become complex.

TOP DEAD CENTER

Top dead center (tdc) on No. 1 cylinder is crankshaft position when No. 1 piston is at its uppermost limit of travel on the compression stroke. This is the datum, or base line, of valve and ignition timing, and it is good practice (and necessary on most overhead cam engines) to disassemble with the crankshaft in this position.

No. 1 is the cylinder most remote from the flywheel, and is obvious on in-line engines. No. 1 on vee engines is the first cylinder on the lead bank (banks are staggered, with one bank half a crankpin length forward of the other).

Most engines have a tdc mark (often indicated as a ''0'' or ''T'') stamped on a plate affixed to the timing-chain cover and indexed to a notch on the crankshaft pulley. Others use a double notch on the pulley and a pointer or casting separation line as the fixed referent. Obsolete and some front-wheel-drive applications time from the flywheel.

Turn the crankshaft in the normal direction of rotation (if you overshoot the mark, go around again—reversing rotation will introduce error and can

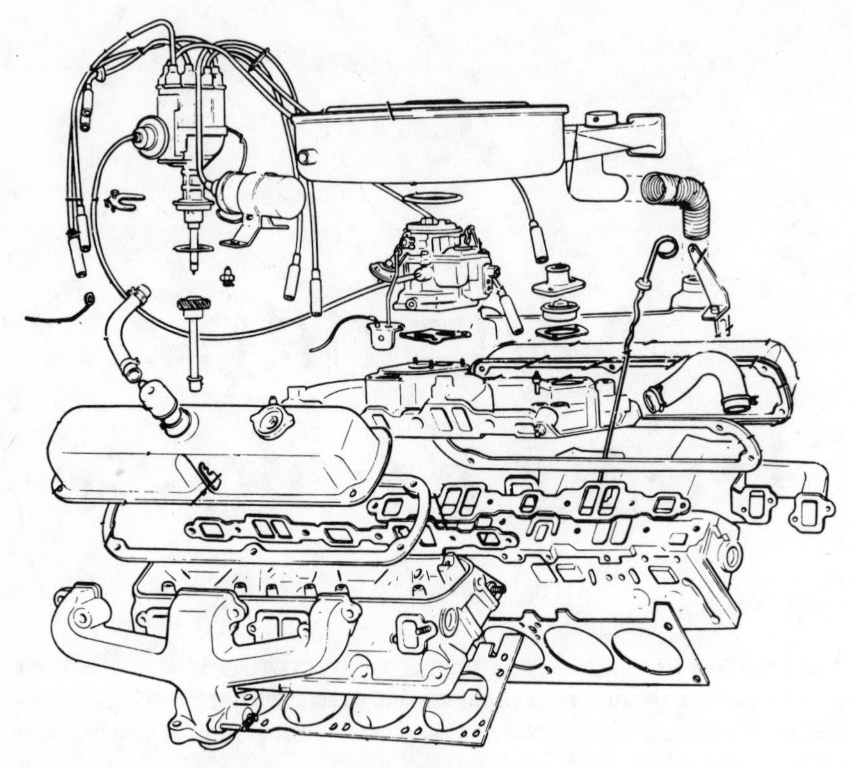

Fig. 4-1. Typical V-8 upper engine configuration.

disengage Gilmer-type timing belts) until tdc marks align and both valves are closed on No. 1 cylinder. As an additional check, verify that the distributor rotor points to No. 1 spark plug terminal (Fig. 4-2).

INTAKE MANIFOLD

The intake manifold is readily accessible on V-type engines; disassembly of in-line engines usually begins with removal of the exhaust manifold (see below). In most cases, the manifold can be detached without unbolting the carburetor/fuel injector or disturbing manifold-to-carburetor hoses. Loosen the hold-down bolts in stages, working in a crisscross fashion from the centermost bolts outward. "Stick 'em" on bolt threads means that the bolt extends into the water jacket. Make note of that fact.

Examine the gasket surfaces for evidence of leakage. A failed gasket may indicate the need for resurfacing the manifold flange. Note any secondary sealing hardware that might be present. For example, early Chevrolet sixes employed short lengths of tubing to make the transition between the intake manifold and cylinder head runners. Current 305 and 350 Chevys use stainless steel clips between adjacent intake manifold runners. Failure to reinstall these parts will result in gas leaks, poor idle, and possibly, detonation.

Some in-line manifolds are bolted to the exhaust manifold through a heater box. In absence of leaks, the wisest approach is to leave these castings assem-

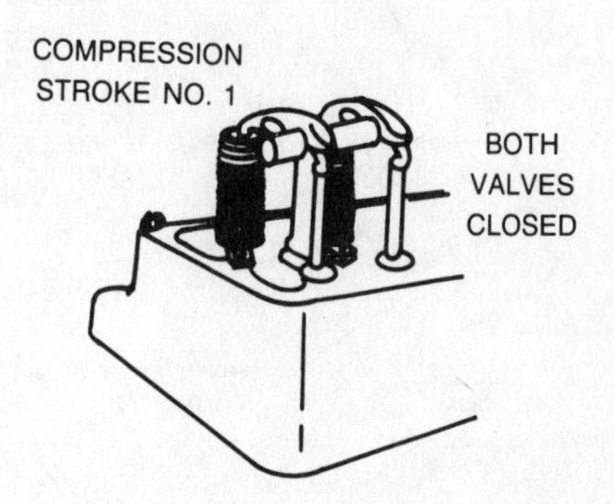

Fig. 4-2. At tdc, both valves on No. 1 cylinder will be closed, the timing pointer will align with the tdc mark, and the rotor will point to No. 1 terminal on the distributor cap. Occasionally, the factory will provide reference marks on the rotor and distributor body to indicate the firing position for No. 1 cylinder.

bled. If separated, one or both castings will require resurfacing and a replacement gasket may be difficult to find. V-type engines may employ a valley cover under the manifold or use the manifold itself as the cover. In the latter instance, the manifold is part of the oiling system, sealed on the ends by rail gaskets and adhesive. Note the gasket layout and make a special effort to clean the underside of the manifold. Splash plates, as used on Ford 352, 390, 406, and 427-CID, collect carburized oil on their undersides and should be removed for cleaning.

DISTRIBUTOR

Figure 4-3 shows a breakdown of a Holley solid-state distributor, mechanically similar to an earlier point-and-condenser version. Ideally, the distributor should be run up on a Sun machine: in practice, most mechanics merely check shaft side play (perceptible wobble is grounds for replacement), cam surface finish (scores or roughness mean rapid rubbing-block wear and loss of point adjustment), and, once the engine starts, gross changes in dwell (which can indicate stator plate wear). The traditional test for the vacuum advance unit is

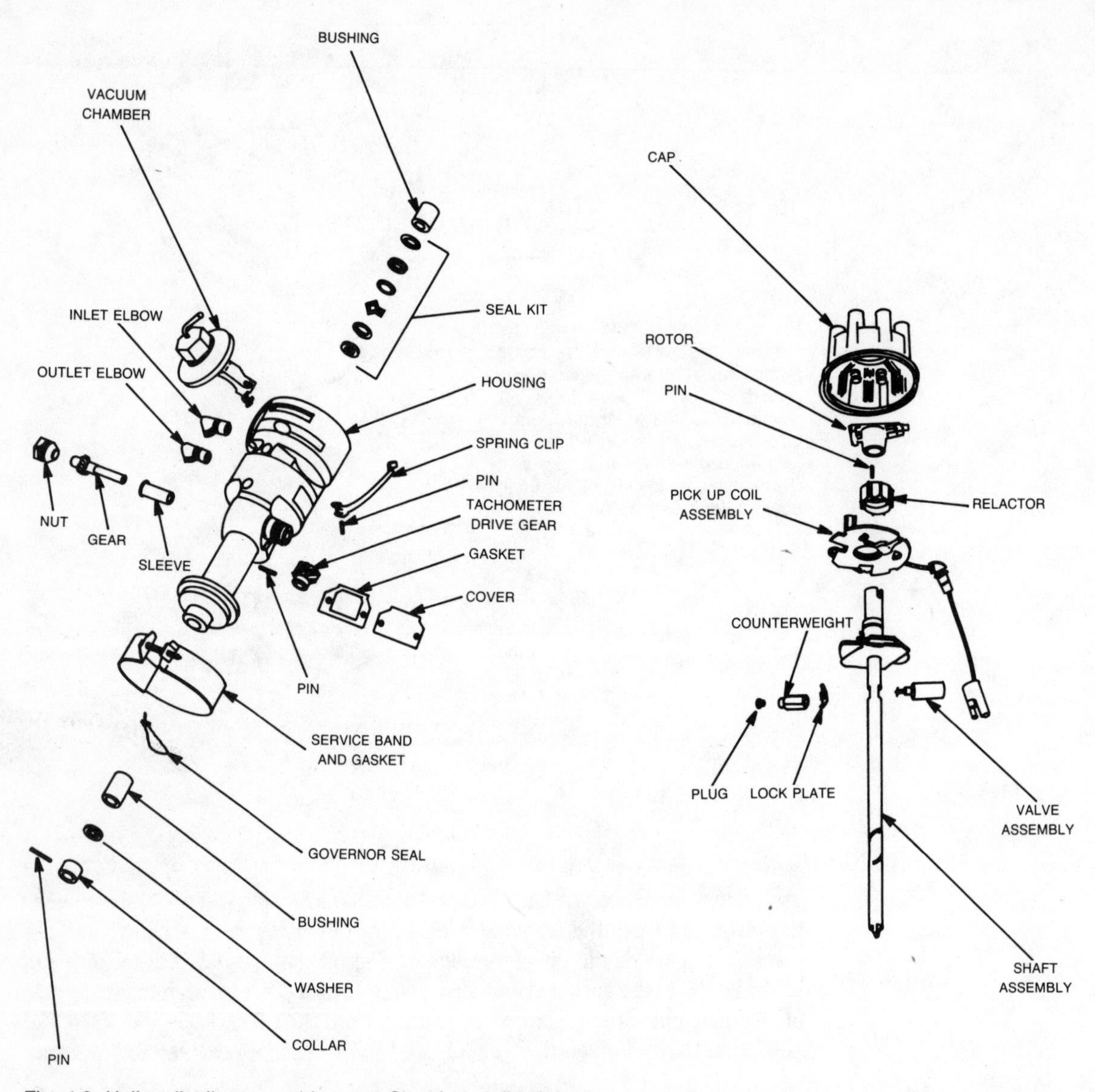

Fig. 4-3. Holley distributor used in some Chrysler applications.

to fully advance the distributor and seal the vacuum port with thumb pressure. If the diaphragm is sound, the stator plate will remain in the full advanced position.

Note the lay of the spark plug wires. Neat, parallel wiring is not necessarily advantageous and can cause cross-firing (Fig. 4-4). If wires have previously been distributed, consult the drawings in the service manual.

Mark No. 1 spark plug wire and its terminal on the distributor cap. You may wish to mark all wires, but make certain that you know which cap tower serves No. 1.

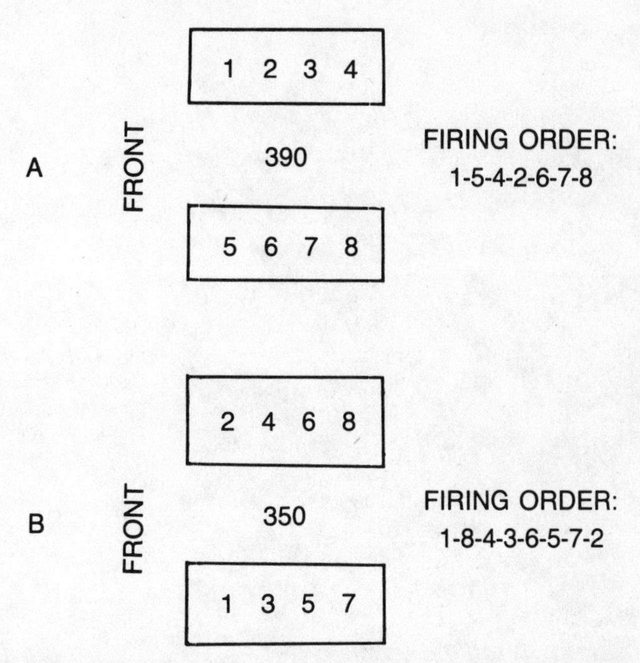

Fig. 4-4. Some engines are more sensitive to cross-fire—an inductive pickup of voltage between spark plug leads—than others. The Ford 390 and derivatives fire cylinders 7 and 8 sequentially ("A"). The natural tendency would be to route the spark plug wires for these two cylinders in parallel runs. But that would be a mistake: wires to these and other cylinders should be crossed over each other and separated as much as possible. The Chevrolet 350 and its kin ("B") are configured with sequentially firing cylinders on opposite banks and, thus, are less liable to cross-fire because of wire routing.

Determine the normal direction of distributor rotation from the manual or by observing the action of the vacuum advance mechanism. When activated, the mechanism rotates the stator plate against the normal direction of cam rotation. (This applies only to low-tech distributors—some units work in both directions, and others employ vacuum as a means of retarding the spark during turbo boost.)

Now, with the preliminaries out of the way, determine the positions of the rotor and distributor body with No. 1 cylinder at tdc on the compression stroke.

1. Verify that No. 1 cylinder is, in fact, at tdc and that its valves are closed.
2. Disconnect and label the wires going to the distributor body.
3. If you have not already done so, remove the distributor cap.
4. Using a Magic Marker, mark the distributor housing directly under the firing tip of rotor (Fig. 4-5). This mark should fall under No. 1 spark plug tower on the cap.
5. Note and sketch the position of some salient feature of the distributor housing. For example, "vacuum advance diaphragm at 9 o'clock." You may wish to scribe reference marks on the distributor body and engine block.

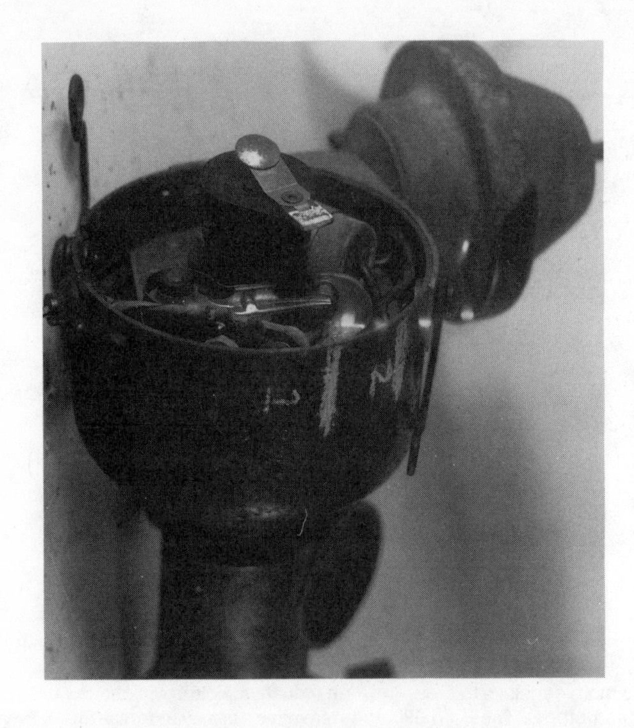

Fig. 4-5. With No. 1 cylinder at tdc, reference the position of the rotor to the distributor housing.

6. Remove the clamp bolt that secures the distributor to the engine.
7. Slowly lift the distributor out of its boss. Note that a gear-driven rotor will turn a few degrees as it lifts clear of the camshaft. In this event make a second mark on the distributor body corresponding to the final position of the rotor (Fig. 4-6). When you come to install the distributor, the housing will be aligned as indicated in Step 5 and the rotor will be turned to the second mark. Once home, the rotor should be on the initial mark made in Step 4.

EXHAUST MANIFOLD

Remove the exhaust manifold(s), again working from the center bolts outward. Take your time with these bolts; breaking one might cost a Heli-Coil job (Fig. 4-7). Some relatively recent exhaust manifolds use no gasket but, to be on the safe side, should be assembled with an aftermarket gasket that may not be included with the standard gasket set.

Repeated heating and cooling cycles can warp and corrode the gasket surface. More than .004-in. runout is the reasonable limit. Your machinist can check this or you can do it yourself, using a precision straightedge and a set of feeler gauges. A similar operation is described and illustrated in the next chapter with reference to cylinder heads. The corrective machine work is quite inexpensive.

Cracks are the result of stress corrosion, water splash, and/or chronic overheating. Not much can be done about water splash, but overheating usually

Fig. 4-6. Remove the hold-down clamp and slowly lift the distributor. The rotor, if driven by a spiral gear, will turn. Mark the final rotor position on the distributor body.

has some correctable cause. Check that the heat riser valve (the butterfly valve sometimes present at the manifold collector) works freely or, if frozen, is in the open position, and check also for restrictions in the exhaust system. This condition can usually be detected with a vacuum test. A slow drop of the gauge needle at steady throttle openings indicates excessive back pressure. Often you can detect by ear any separation of the inner exhaust pipe from the outer, with the engine running at 3,000 rpm or so. Failure usually occurs on the downstream side of the Y-collector.

Overheating might also reflect severe operating conditions. Chrysler 318 van and motor-home engines seem to have this problem. A free-flowing exhaust system and steel-reinforced manifold gaskets help.

Stress corrosion inevitably results from the heating and cooling cycles imposed on the manifold. Shot peening dramatically reduces stress corrosion and, if no other problems are present, more than doubles manifold life.

Local cracks can be brazed or heliarcked, although a wrecking-yard replacement part is often the better solution.

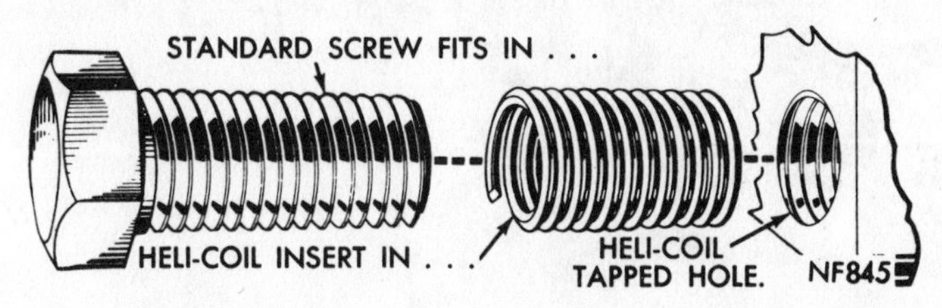

Fig. 4-7. Heli-Coil inserts are used for permanent repair of damaged or worn threads. Essentially, the repair consists of drilling out the damaged threads, tapping special Heli-Coil threads, and installing the insert. (Courtesy Chrysler Corp.)

VALVE COVER

Remove the valve cover(s), checking for evidence of oil leaks. One of the worst offenders is the plastic cover used on 1981-82 AMC 258 engines (Fig. 4-8). RTV sealants, even exotic products such as General Motors type B (PN 105297), have failed to control these leaks. The best solution, short of fabricating another valve cover, is to use a comformable pre-cut gasket, such as the Fel-Pro VS-50244 R, which is the OEM part. Late production (mid-1986) 258 valve covers are sealed with a bead of RTV that should be left in place and reused. Small cracks can be repaired with high temperature (red) RTV.

Metal valve covers can be sent out with the other engine parts for chemical cleaning.

Make a note to replace the positive crankcase ventilation (PCV) valve. Inspect the valve-mounting grommet (usually made of plastic, and brittle as glass) and the hoses for leaks. If serious amounts of oil are present, swab the hoses with a rag and coat-hanger wire. The splash baffle, spot-welded to the inner side of the valve cover, has been known to come adrift on some GM engines.

CYLINDER HEAD

Cylinder heads are classed by valve and cam arrangement. In the overhead valve (ohv) configuration, the valves are operated via pushrods from a block-mounted camshaft. In the overhead cam (ohc) configuration, the camshaft is mounted in the cylinder head above the valves.

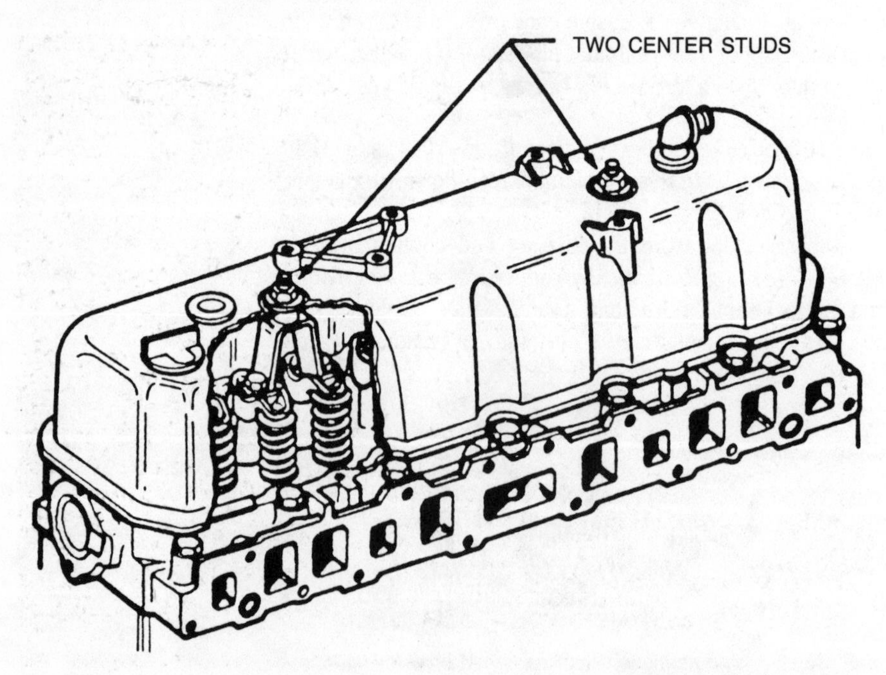

Fig. 4-8. AMC plastic valve cover, used on certain 258 engines, suffers from distortion and cracking. Lack of perimeter bolts exacerbates the problem. (Courtesy Fel-Pro Inc.)

Overhead Cam

Most overhead cam (ohc) passenger car and light truck engines employ a single cam (or one cam per cylinder bank) that acts on the valves through rocker arms that, in these applications, are also known as cam followers or "fingers" (Fig. 4-9). Ohc engines incorporated some provision for referencing the position of the camshaft relative to the crankshaft. Make certain you understand the significance of these marks before disengaging the cam drive chain or belt.

Belt. With the No. 1 piston at tdc on the compression stroke, remove the belt cover(s), disengage the belt tensioner, and slip the belt off its sprockets. The belt should be renewed as insurance, and especially on engines such as the first-series Ford 98. Belt failure on these and other non-freewheeling engines will send the valves crashing into the pistons.

Chain. The drive chain is disengaged from the camshaft by removing the camshaft sprocket, per the instruction manual for the engine at hand. The chain remains in mesh with the crankshaft, or lower, sprocket until the timing cover is removed. However, there is one problem that service manuals fail to address: most Japanese engines are timed by indexing a punch mark on the camshaft sprocket with a single link or adjacent pair of colored links on the drive chain (Fig. 4-10). The camshaft mark and colored link(s) align when No. 1 piston is at tdc on the compression stroke. While not visible at this stage of disassembly, another colored link is aligned with a mark on the crankshaft sprocket.

Part of the difficulty arises because the color coding is oil-soluble and washes off in service. The real dimension of the problem becomes clear when it is realized

Fig. 4-9. Cam followers function as rocker arms in single-overhead camshaft engines. The followers may be mechanical (note the adjustment screws in this example) or may incorporate hydraulic valve lash adjusters.

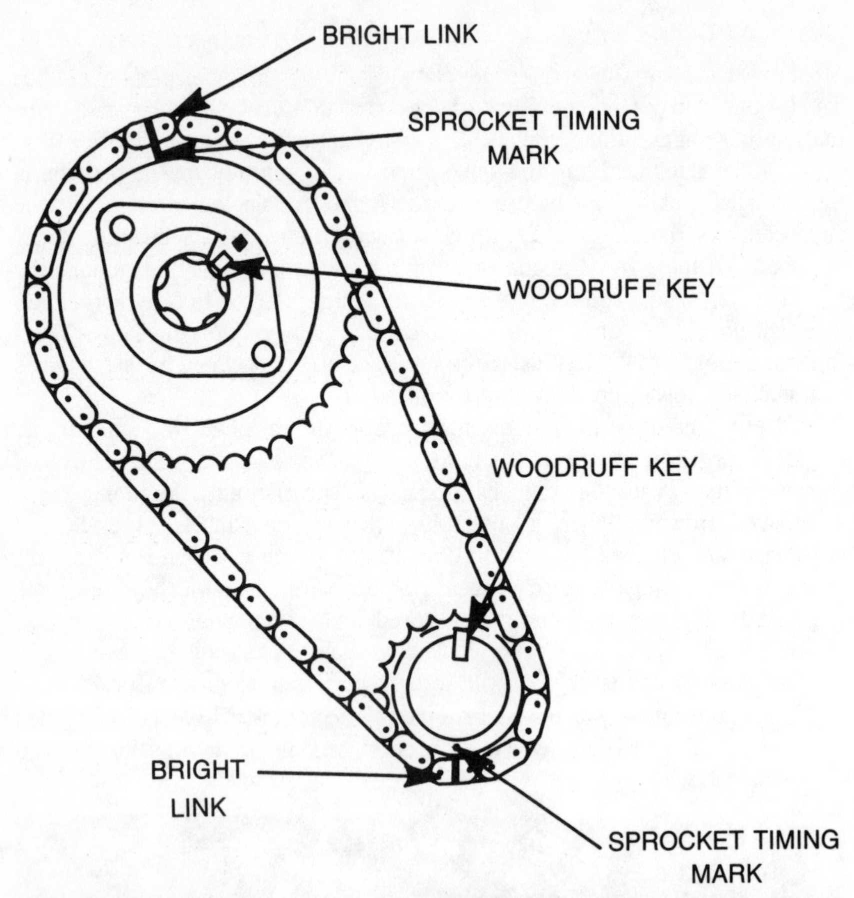

Fig. 4-10. Japanese practice is to color chain links adjacent to marks on cam and crankshaft sprockets. As explained in the text, the distinctive links are a convenience—the real criterion is the number of links between sprocket timing marks, which in this case is 15 on the short side of the chain.

that aftermarket chain manufacturers may omit the color code on the chain-to-camshaft link(s).

While it is possible to time an unmarked camshaft, the process is tedious and not without risk. It is far better to make your own timing marks while the engine is still together and timed as the factory intended. Follow this procedure:

1. Set No. 1 piston at tdc on the compression stroke as explained above.
2. Look for the mark on the camshaft sprocket. The colored link or pair of links should be opposite that mark. (Two links are marked when the camshaft mark falls between them; one link is colored if the camshaft mark falls at the center of that link.)
3. If the link is indistinguishable from the others, wash the oil off and, using a Q-tip, paint the link(s). This is the chain-to-camshaft mark.
4. One critical step remains: upon disassembly, count the number of links from the marked link(s) to the link adjacent to the crankshaft sprocket

mark. For example, let us suppose that 15 links on short side of the chain separate the links adjacent to the crankshaft and camshaft marks. The new chain will then be assembled with the same link spacing.

Much trouble can be avoided if the whole drive mechanism—chain, camshaft, and crankshaft sprockets, hydraulic tensioner, and chain guide rails—are replaced while the engine is apart.

Camshaft/Follower Inspection. Examine the camshaft and follower tips, which ride against the camshaft, for scores and grooving. A scored camshaft lobe means the cam and the affected follower must be replaced. Ideally, all followers are replaced whenever the camshaft is renewed. Where applicable, inspect the follower support shaft and follower bearings, both of which wear at the side nearest the head.

Premature camshaft and follower failure is characteristic of most Japanese and some European engines. It has been suggested that the combination of chilled iron camshafts running with mechanical lifters is lethal. Tremendous forces are generated when the lifters are allowed to run loose, as they usually are.

In addition, the ohc configuration places the cam at the far end of the oiling circuit (Fig. 4-11 is typical). At this remove from the pump, small-diameter oil ports, including those that may be drilled in the camshaft itself, tend to clog.

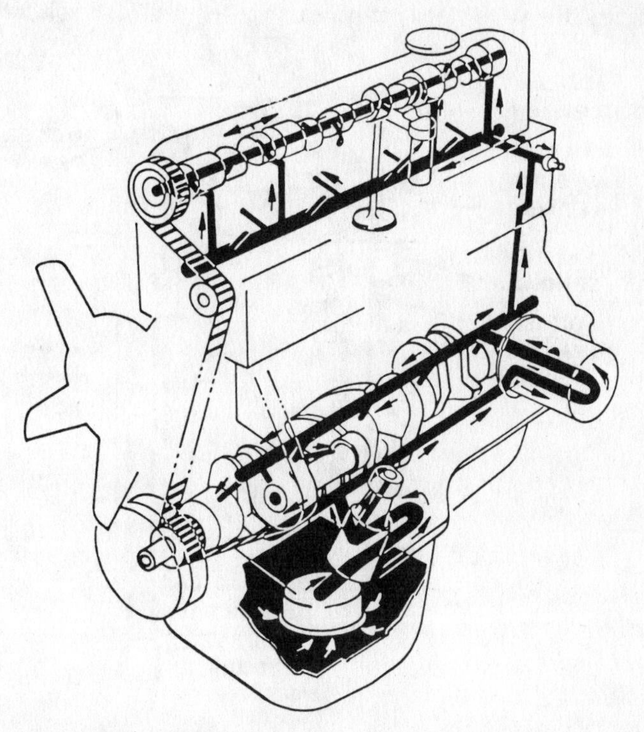

Fig. 4-11. Understanding the oiling circuit is never more critical than for ohc engines. This particular design employs a hollow cam and individual oil ports—with tiny metering holes—at the lobes. (Courtesy Sealed Power Corp.)

Do not extract the camshaft; most machinists prefer to handle this critical aspect of the work themselves. In the case of Pinto and Vega engines, special tools are required. Some engines feature replaceable cam bearings, which are either full-circle bearings (requiring a special tool to install) or split inserts, similar to conn-rod bearings. Other engines, including BMW, most late Toyota production, and the new GM Quad 4, run the cam against head metal. Serious wear means a new, or a good used, head casting.

Overhead Valve

Figure 4-12 illustrates a pedestal-type overhead valve (ohv) arrangement, typically run in conjunction with hydraulic lifters. Stamped-steel rocker arms pivot on individual ball nut and stud assemblies. The assembly illustrated is adjustable, in that the ball can be raised or lowered to center the lifter piston in its bore. Figure 4-13 shows two variations that use fulcrum pieces, secured by stud or capscrew, in lieu of a ball nut.

The alternative to pedestal rocker arms is to pivot the rockers on shafts. When intake and exhaust valves are arranged in a single plane, one shaft suffices. More efficient combustion chambers require separate intake and exhaust rocker shafts. The rocker shaft doubles as an oil gallery, distributing oil to the rocker arms, which are usually steel castings or forgings. Screw-type adjusters, acting directly against the valve stems, provide lash adjustment. But adjustment is not

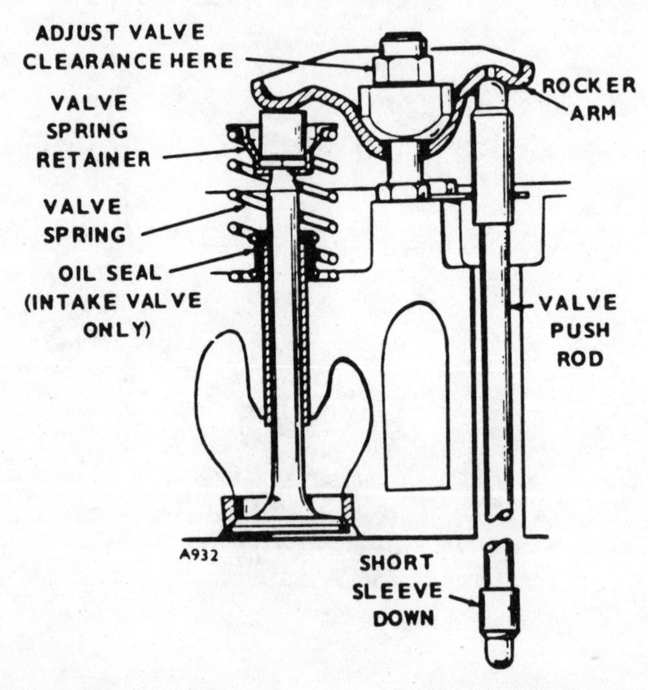

Fig. 4-12. A pedestal-type rocker arm, using a locknut for lash adjustment. Note that pushrods are not reversible end for end. (Courtesy Onan)

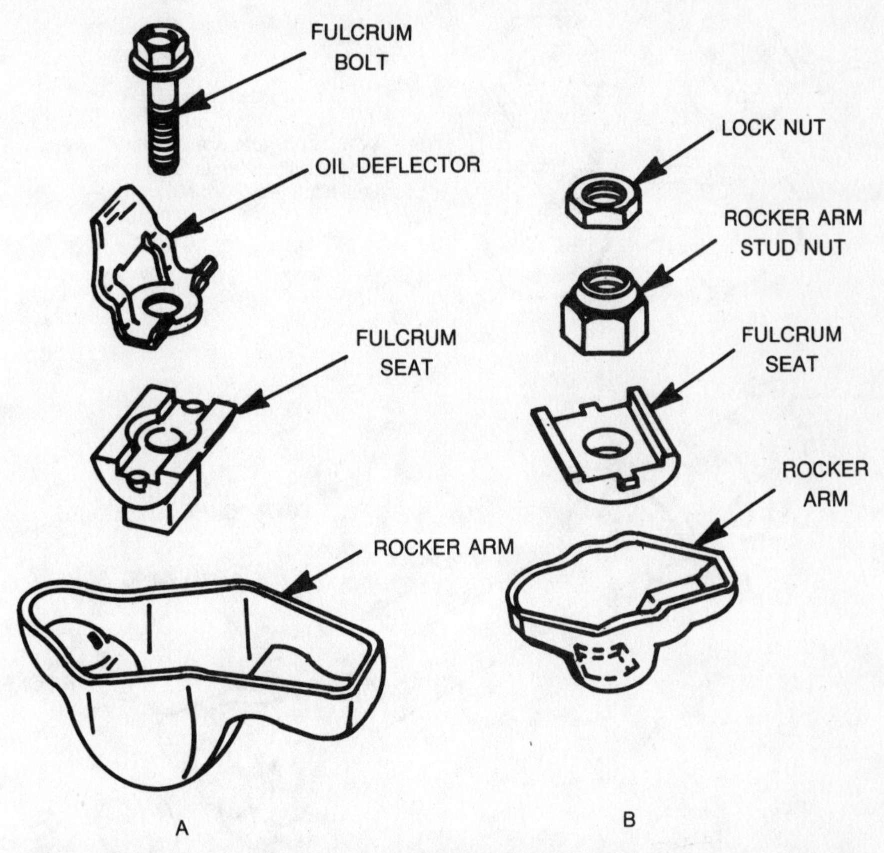

Fig. 4-13. Fulcrum seats may be secured by nonadjustable capscrews ("A"), or, in certain Ford engines with mechanical lifters, by studs and adjustable nuts ("B").

always possible or convenient. Chrysler V-8s are nonadjustable. Slant-Six engines from the same company initially used mechanical lifters and so had provision for adjustment. When hydraulic lifters were introduced on these engines in 1981, the adjustable rockers were eliminated.

DISASSEMBLY

Two different approaches apply here.

Pedestal-Type. Back off the rocker hold-down nuts far enough to slip the cupped ends of the rockers off the pushrods. Stud nuts, which bear against the balls or fulcrum pieces, tend to fret their threads and wear at the contact faces (Fig. 4-14). Conscientious mechanics replace these nuts as a matter of routine.

Shaft-Type. Remove as an assembly with the shaft (Fig. 4-15). Note that rocker-shaft fasteners are tensioned by the springs of those valves that are open. Work from the centermost fasteners outward, equalizing tension along the whole length of the shaft. Shaft bolts may not be identical: one can be grooved or drilled for oil, and others can extend into the water jacket, as discussed below.

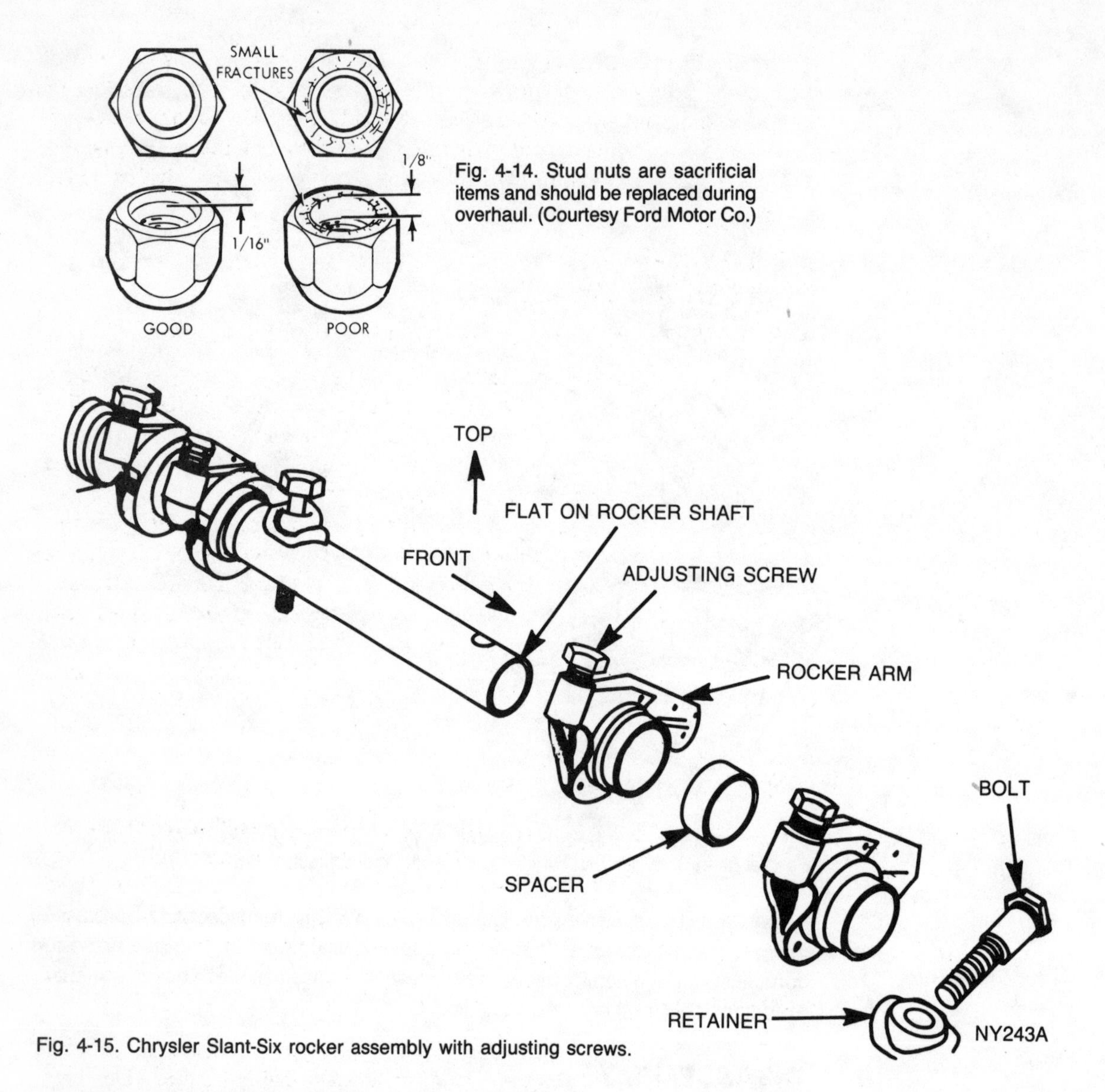

Fig. 4-14. Stud nuts are sacrificial items and should be replaced during overhaul. (Courtesy Ford Motor Co.)

Fig. 4-15. Chrysler Slant-Six rocker assembly with adjusting screws.

INSPECTION

Remove one pushrod at a time, placing it upright in a numbered rack. (A two-by-four, with holes corresponding to the number of pushrods and marked to show the front of the engine, will serve.) The business of racking the pushrods is important, both on the general principle that used parts should be assembled with their mates and because all pushrods may not be the same length. Some Ford production mixes standard length pushrods together with 0.60-in. over and under rods. It is also possible to install a pushrod upside down, with unfortunate results. Both ends of pushrods used with pressed-steel rocker arms terminate in balls that, as often as not, have different diameters.

Clean each pushrod with solvent and, where applicable, blow out the ID with compressed air. Check for abnormal wear on pushrod ends and sides. In the absence of factory specs, maximum allowable deflection can be taken .025 in. at the center of the pushrod. This can be determined by rolling the pushrod on a piece of optically flat plate glass and measuring the "bow" with a feeler gauge.

PEDESTAL-TYPE ROCKER ASSEMBLIES

Clean and inspect with particular attention to the following parts:

☐ Studs—check for nicks, thread wear, straightness, and separation from the head. The aftermarket supplies threaded studs as a more reliable alternative to the pressed-in type. A defective pressed-in stud can be extracted with the aid of a wrench socket or a stack of washers. Slip the spacer over the stud and screw down the nut, adding washers as necessary.

☐ Rocker fulcrum and ball—reject both parts if either is deeply scored, discolored, or shows surface cracks. One of the difficulties with this type of rocker is that the ball and rocker wear in to an oil-tight seal. Grooved balls, sometimes OEM, most often aftermarket, eliminate the problem. Used balls or fulcrum pieces must be stored with their rockers and assembled as found.

☐ Rocker tips—look for evidence of impact damage that could point to a failed hydraulic lifter and possible valve tip/valve guide/pushrod damage.

SHAFT-TYPE ROCKER ASSEMBLIES

Rockers should be removed from their shafts—an operation that requires care. Many engines employ left and right-hand rockers, and rockers may vary between engine banks. Make note of the location of spacers and springs. Most rocker-arm assemblies are secured by a C-ring or cotter pin; Buick uses throwaway nylon locating pins that are disassembled with a cold chisel. If no factory mark is present, clearly mark the forward end of the shaft and note that the oil holes are down, toward the head. Critical parts are:

- ☐ Adjustment screw—look for loose or stripped threads, and severe mushrooming at the tip, which can indicate that the surface hardness has worn through.
- ☐ Rocker bearing—scores, severe wear on underside. Some rockers use pressed-in bushings, others ride on iron.
- ☐ Rocker shaft—same as above.
- ☐ Oil-related failure—once again, make certain you understand the oiling circuit. Clean oil ports with a drill bit.

SPARK PLUGS

Removing the spark plugs is a routine operation, which would hardly be worth mentioning were it not for the danger of stripping the outermost threads. If carbon deposits on the lower threads bind the plug, apply penetrating oil and alternately tighten and loosen until the plug can be coaxed out by hand. Using a socket wrench at the end of thread travel can strip the threads.

Combustion deposits should be fairly uniform on all spark plugs. Sharp variations in spark plug color and deposit texture indicate localized problems that should be investigated. For example, clusters of black carbon on the firing tip of one spark plug and relatively clean plugs in other cylinders mean a localized oil-burning problem; white deposits can mean an intake-manifold or vacuum-system air leak on the affected cylinder; oily deposits on spark plugs from adjacent cylinders may indicate a blown head gasket. The truth of these conjectures will be known once the head is removed.

HEAD BOLTS

There are four major concerns about cylinder head bolts:

☐ Can the bolt be reused? Traditional capscrews are subject to twisting, thread damage, and fatigue cracks (usually just below the head). Some of the newer torque-to-yield bolts are throwaways, intended to be used once and discarded. Examples include 1984 Mercury 98 non-turbo bolts and some 1985 GM applications. High-grade gasket sets provide the necessary bolts and torque instructions.

☐ Bolt length? A V-8 engine may employ three different head bolt lengths. Mixing these bolts invites stripped threads and/or a broken water jacket.

☐ Which, if any, bolts enter the water jacket? Some head bolts—all of them on small-block Chevy V-8s—may be wetted (Fig. 4-16). These bolts must be assembled with an appropriate sealant to keep water out of the oil supply. It is equally important not to use sealant on bolts that go into blind holes. A service manual will identify the critical bolts; otherwise, look for traces of previously used sealant and/or corrosion on lower bolt threads. Use a flashlight to determine which bolt holes are drilled through.

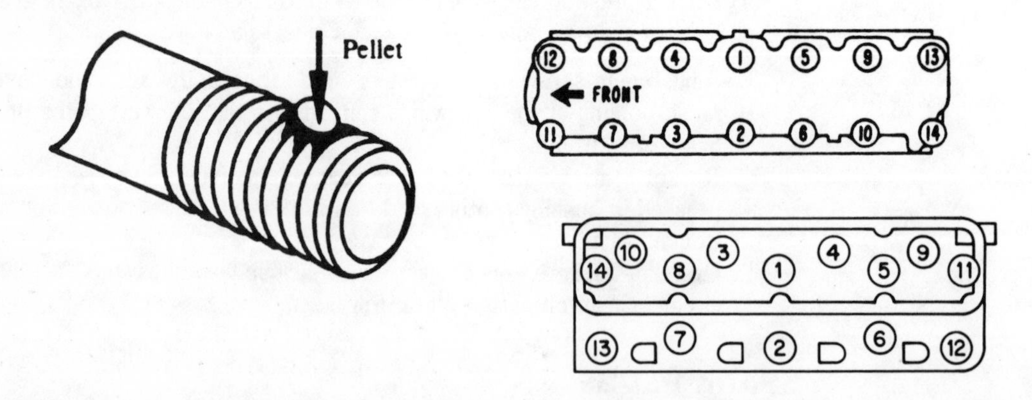

Fig. 4-16. Thin-wall castings mean that cylinder head, rocker arm, and water pump bolts may extend into the water jacket. These bolts must be sealed. (Courtesy Fel-Pro Inc.)

Clean the threads of reusable bolts with a wire brush and chase the bolt hole threads with an appropriately sized tap.

In the absence of specific instructions from the manufacturer, loosen the head bolts in two or three stages, following the assembly torque sequence (see Chapter 7). This precaution is critical for aluminum heads, which are easily warped.

Note that Japanese practice is to secure the timing gear cover to the head with small, camouflaged bolts. These bolts must be removed before the head bolts are loosened. Otherwise, the head warps.

The breakaway torque for all bolts should be the same. Go easy on the stubborn ones, alternately tightening a degree or two and backing it off the same amount. Penetrating oil may help, provided it goes where directed. Do not force the bolt past the yield point, when torque requirements suddenly relax. Sudden loss of torque can mean that the bolt is verging on failure. There is a special case involving aluminum head castings; on rare occasions one encounters a bolt that unthreads easily enough but remains captive in the head. The problem can be traced to accumulations of aluminum oxide (or aluminum "rust") on the bolt OD. Penetrating oil will only make things worse. If you are reasonably certain that the threads are, in fact, clear of the block, remove the head and drive the bolt out with a punch or press.

HEAD REMOVAL

At this point, the head has been disassembled to the valves (or valves and overhead camshaft), which is about as far as the occasional mechanic ought to go. All that remains is to detach the head from the block. The head should lift off, perhaps binding slightly during the initial quarter-inch of travel until the locating dowels are cleared. A few sharp raps with a soft-faced mallet may be required to break the gasket seal. If this does not work, check for an overlooked head bolt. Do not attempt to pry the head off with a screwdriver.

You may encounter a cylinder head that appears to have grown in place. The problem is corrosion, often exacerbated by liberal use of gasket cement. These heads can be unstuck by spinning the engine with the starter with the spark plugs in place. The H-bomb of head removal techniques is to reassemble the valve gear and start the engine.

INSPECTION

Support the head, face up, on 2-by-4s and examine the carbon deposits in the combustion chambers. We hope to find uniformity. A suspiciously clean chamber, which may show traces of coolant, means a head gasket or casting leak. Noticeably oily chambers mean piston-ring, valve-seal, or, on certain V-type engines, intake-manifold-gasket failure. Breaks in the head gasket or flow channels across gasket surfaces mean leaks and raise the possibility of a warped head. Air leaks or local overheating tends to bleach out the deposits. Such abnormalities should be noted and reported to the machinist.

Save the old head gasket for comparison with the replacement part.

5
CHAPTER

Head Servicing and Assembly

HEAD WORK IS THE PRESERVE OF THE MACHINIST WHO, ONE TRUSTS, HAS THE necessary skills and tools. But it is to the owner's advantage to be more than a silent partner in this endeavor. He should be able to communicate with his machinist, and should help decide how far the work will go, what trade-offs will be made between long-term durability and immediate cost. Auto mechanics has become something like medical technology, in that the cost of repair can outweigh the practical use. In addition, the owner must be able to evaluate the finished product.

CLEANING

Most shops begin by steam-cleaning the head, together with its assembled valves. Iron heads are then "hot-tanked" in a strong caustic solution, heated to near-boiling temperature. This process dissolves grease, paint, rust in the coolant passages, and does a fair job on carbon. No single and relatively nontoxic chemical works for aluminum. The part is first degreased, rinsed, and then may be immersed in a solvent to remove paint and carbon. This is followed by a rinsing cycle and immersion in a mild acid bath to remove scale from the coolant passages. What carbon remains is removed by hand. When deposits are heavy or when appearance is a factor, head castings may be dressed out with glass-bead or light sand blasting.

VALVE DISASSEMBLY

Valves are then removed by compressing the springs just far enough to disengage the valve locks, or keys (Fig. 5-1). Mushroomed tips that do not pass easily

Fig. 5-1. Many valve spring compressors are in the form of a lever that reacts against the head to collapse the spring. Others compress the spring with a clamping action.

through the valve guides should be dressed to size with a stone. Valve assemblies should be kept together as wear-matched sets and replaced in their original order.

CRACKS

Heads "work," flex in service, and are subject to extreme thermal gradients. Most cracks occur between the intake and exhaust valve seats, but no part of the head casting is immune.

Detection

Visual inspection must be supplemented by more sophisticated techniques.

Magnetic Particle Testing. Generically called Magnafluxing, this inspection technique is the fastest and least expensive for ferrous metals (Fig. 5-2). It does not work on aluminum. In basic outline, the procedure goes like this:

- ☐ The part to be tested is subjected to a powerful magnetic field. The head becomes, at least temporarily, a magnet.
- ☐ Breaks in the magnetic circuit, represented in this case by a crack, become tiny magnetic poles. One edge of the crack is north pole; the other; the south pole.
- ☐ Finely divided iron particles, sprinkled over the head, congregate at the poles to outline the otherwise invisible crack.

Penetrant Dye. This method, which reveals cracks through capillary action, may be used on aluminum and other nonferrous metals:

- ☐ The head is thoroughly cleaned, and the penetrant, usually packaged in aerosol cans, is applied.
- ☐ Within a few minutes, the dye penetrates into any cracks that may be present. The surplus is wiped off, and
- ☐ A chemical developer is applied, which draws the dye out of the cracks.

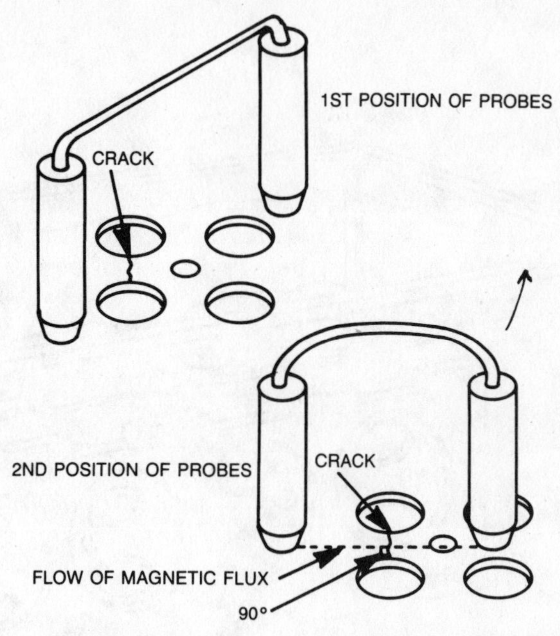

Fig. 5-2. Cracks show up best when they are at 90 degrees to the test probes. A Seal-Lock unit is shown.

These and other indirect methods of crack detection are useful, but so far as the water jacket is concerned, lack the persuasiveness of the old-fashioned pressure test:

- ☐ All water jacket openings are sealed with gasketed plates. An air line connects to one of the plates.
- ☐ The head is immersed in water, and compressed air is applied to the water jacket. Bubbles indicate a leak (Fig. 5-3).

A few shops (usually specializing in diesel service) go further, and run the test with the casting immersed in hot (200 degrees F.) water. This expands the metal to open cracks that might otherwise remain dormant until the engine reaches operating temperature.

Repairs

Cracks can be repaired by welding—usually acetylene for cast iron and tungsten inert gas (TIG) for aluminum. Iron heads might also be salvaged by cold-pinning or "sewing" the crack. Figure 5-4 illustrates a clever wrinkle, which amounts to toenailing a crack in iron. A special fixture is used to three tangent holes at intervals of about 1 in. across the crack. The holes are splayed so that, if drilled deeply enough, they would meet at a common apex. Seal-Locks—siamesed pegs with angles that match the predrilled holes—are driven into place. The angled pegs pull the sides of the crack together. Areas between Seal-Locks are filled with tapered and threaded plugs. The repair surface is then peened over with an air hammer and dressed with a grinder.

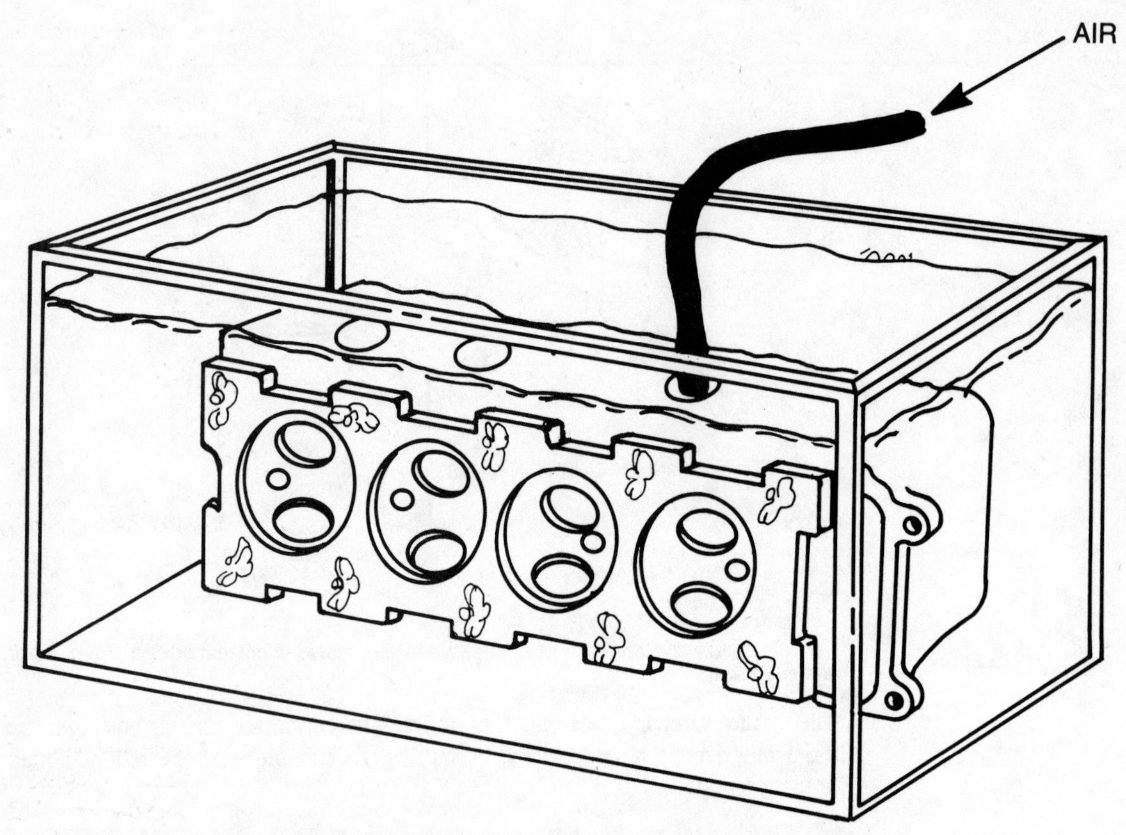

Fig. 5-3. The head should be pressurized at about 50 psi and immersed in water. A stream of bubbles will lead you directly to the cracks . . . even those hidden far up in ports or around valve seats. (Courtesy T. Hoff Mfg.)

Done right, crack repair is permanent, but you may wish to investigate the purchase of another *guaranteed* head from a wrecking yard. Make certain that the head does not merely bolt up, but that valve sizes, gasket holes, and combustion chamber volume are identical to the original.

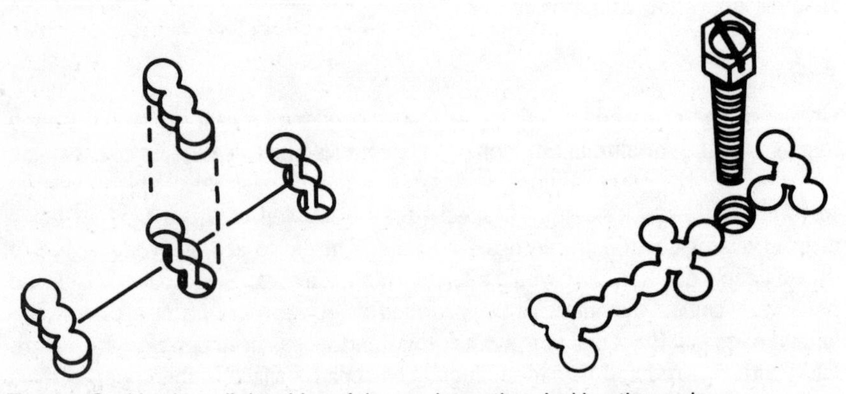

Fig. 5-4. Seal-Locks pull the sides of the crack together, locking the wedges.

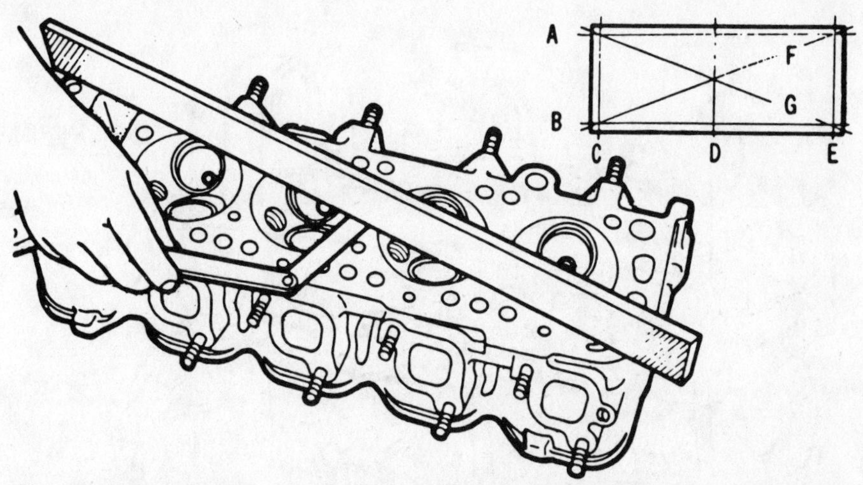

Fig. 5-5. Feeler gauges and a machinist's straightedge are used to determine cylinder head warp. The procedure should be repeated on ohc camshaft bearing pedestals. (Courtesy Chrysler Corp.)

STRAIGHTENING

Figure 5-5 illustrates the technique for determining cylinder-head warp. Most shops are satisfied if the head is true to 0.00075 times the span length. For example, a 12-inch-long head with less than .009-in. distortion would be acceptable by this formula. Manufacturers are more rigorous, particularly when an overhead cam is involved. Nissan holds 240Z heads to a minuscule 0.0039 in., or 0.0019 in. more than the original work limit. The gasket manufacturer Fel-Pro, which has a proprietary interest in these matters, suggests that distortion in three-cylinder heads (i.e., V-6) be limited to 0.003 in. on the long dimension. Four-cylinder heads are allowed 0.004 in., six-cylinder heads 0.006 in.

Small misalignments can be corrected by either milling or grinding the gasket surface. This usually suffices for iron heads. Aluminum heads, for reasons discussed below, often need straightening and may require other corrective work before they can be resurfaced.

Most recently designed engines use an aluminum ohc head and a cast iron block. Like all metals, aluminum and cast iron expand when heated. Movement is in all planes, but is greatest in the long dimension. The rate of thermal expansion for aluminum is four times that of iron. Head bolts restrict lateral movement and the head bows upward, lifting 0.040 in. or more at the center. Areas under bolts nearest the bend may be deeply imprinted (Fig. 5-6).

Distortion of this magnitude defeats the head gasket and may bend the camshaft on ohc engines. In any event, the center camshaft bearings will be out of vertical alignment with the end bearings. Secondary effects include displacement of valve seats relative to the valve guides.

Heating the underside of the head to about 500 degrees F. usually brings the gasket surface within 0.010 in. of true and well within the range of resurfacing. However, cam alignment may still be a problem. Removable cam bearing pedestals (e.g., Datsun, Mercedes) can be machined and shimmed. Heads with

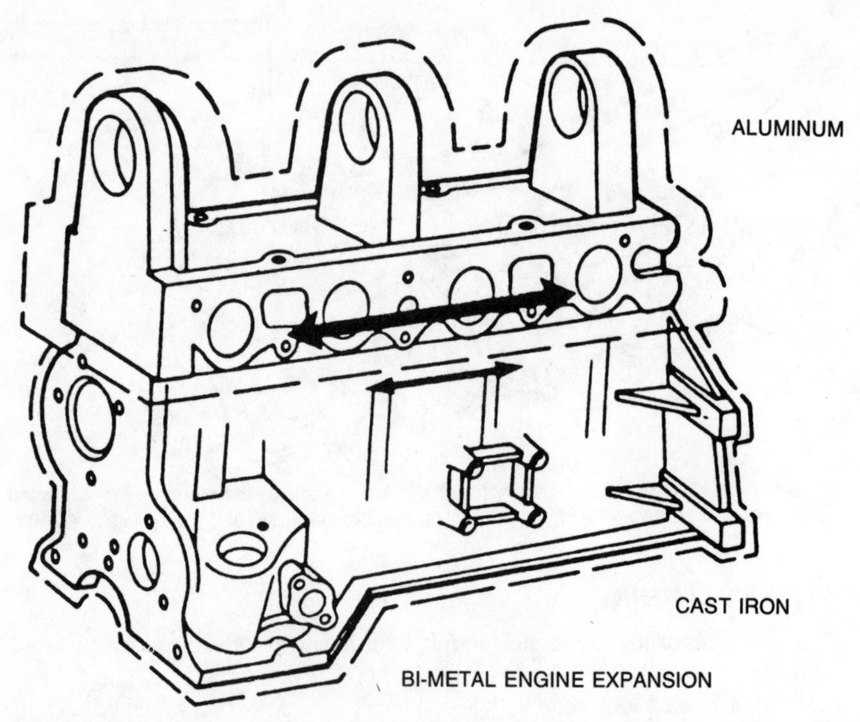

Fig. 5-6. Thermal expansion for an aluminum head is about four times greater than for a cast iron block. When these parts are bolted together, something must give. (Courtesy Fel-Pro Inc.)

split bearing caps (VW, Audi, Mazda-Courtier, some Toyota) can be line-bored or honed. In this case, partial straightening may require a 0.002- to 0.003-in. cut on the center bearing caps, but the outermost bearing bores and camshaft-to-crankshaft distance remain undisturbed. Heads with full-circle integral bearings (such as Ford Pinto) must be straightened or scrapped.

RESURFACING

Resurfacing is a routine operation that removes minor blemishes from the cylinder-head gasket surface and, when pushed further, can compensate for some of the effects of head warp. Light resurfacing is usually done on a blanchard grinder (Fig. 5-7). Heavier cuts of 0.015 in. or more call for a rotary broach (known as a "mill" in the automotive trades) that, when set up correctly, is far more accurate than a grinder. Moving-belt grinders have become popular in recent years, especially in production shops. These machines are relatively inexpensive, require zero set-up time, and produce an unsurpassed finish. But, at least as currently developed, belt grinders, which react off a platen, are less precise than blanchard grinders, and probably should be confined to finish operations.

Stock removal must be held to a minimum. As a rule of thumb, 0.060 in. off the heads will raise the compression ratio one point on a V-8 engine. That

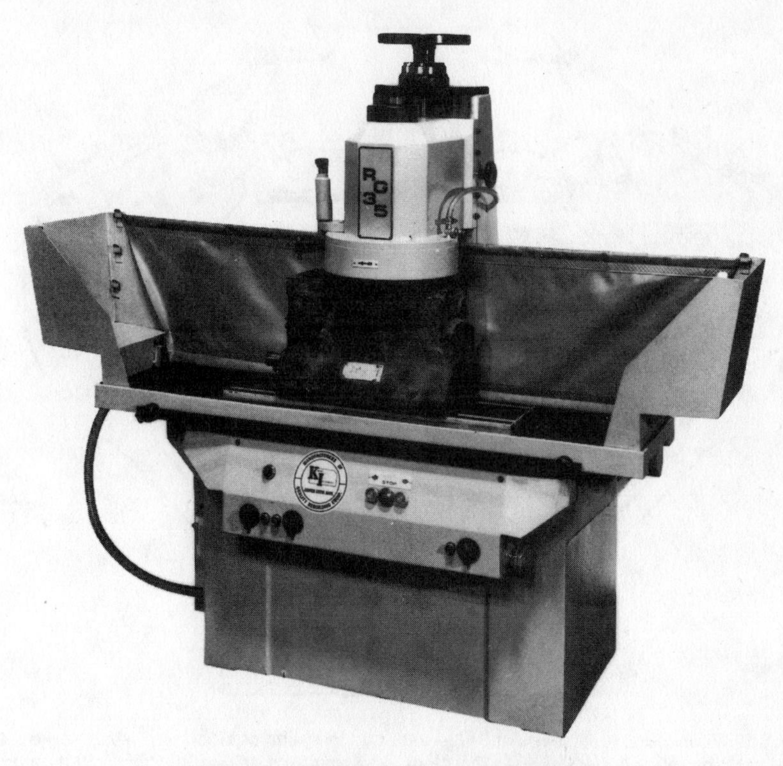

RG35 SURFACE GRINDER

Fig. 5-7. A blanchard grinder is the workhorse of automotive machine shops, used for light-to-moderate head milling, manifold milling, and other jobs where some loss of precision can be tolerated.

is a significant increase and can make the vehicle virtually undrivable. Milling the head or block will lower the rocker arms, causing the valves to open earlier in the rocker-arm stroke. As a result, some of the downward thrust of the rocker appears as a side load on the rocker pivots and valve guides. Wear on these parts will be severe, unless the valve stems are shortened to compensate for loss of metal at the fire deck. Overenthusiastic milling can also lead to piston/valve collision, an event of rather profound significance.

Ohc engines are a special problem. Milling the head retards the camshaft relative to the crank. The valves open late, shifting the power peak toward the high end of the rpm scale. Part-throttle performance suffers and exhaust emissions may be adversely affected. As a general rule, 0.015 in. is the maximum that can be safely removed from an ohc head and/or block.

MANIFOLD MACHINING

Removing metal from the cylinder heads and/or block firedecks on vee engines repositions the intake manifold deeper into the valley formed by the cylinder banks. There is usually enough slope at the manifold flange bolts to accommodate

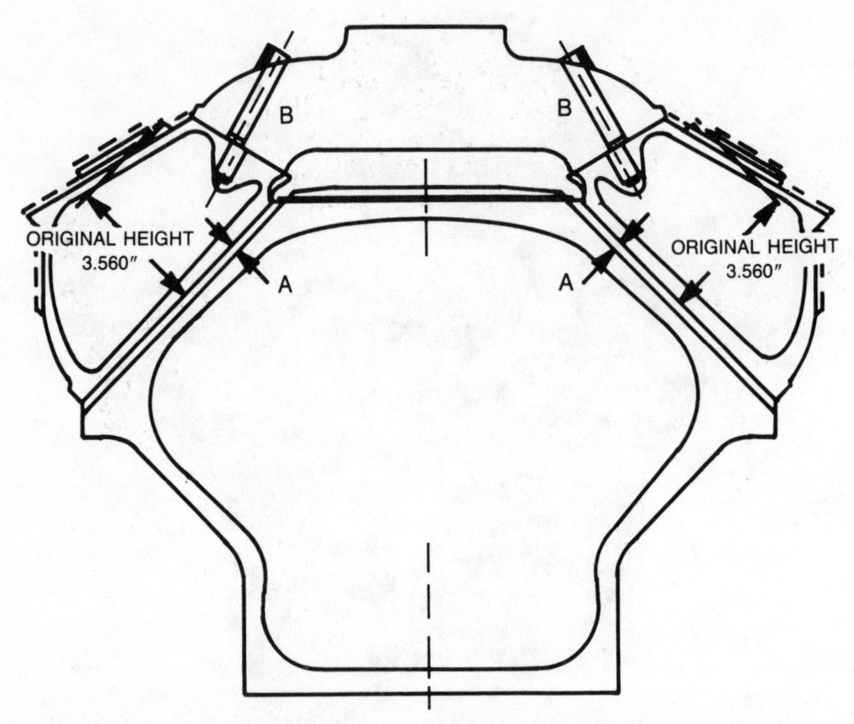

Fig. 5-8. When metal is removed at "A"—either on the head or block side—the intake manifold must be repositioned lower in the block. This requires milling the manifold flanges at "B." The depth of the cut depends upon the initial cut at "A" and upon the angle of the head relative to the block.

light resurfacing cuts, but whenever serious metal has been shaved from either side of the head gasket surfaces—"A" in Fig. 5-8—it is good practice to make a trial assembly of the manifold and gasket. Have the manifold flanges—"B" in the drawing—milled if the bolts that secure the manifold to head fail to center in their respective holes.

Note that the depth of cut depends upon the amount of metal previously removed from the block/head interface, and upon the angle of the head relative to the block. When the manifold serves as the lifter valley cover (e.g., Chevrolet small block), interference can occur at block rails, area "C" in Fig. 5-9. "C" is milled according to the following formula: (developed by Hastings Manufacturing Company):

Metal removed at "C" = Metal removed at "A" × 1.71. Thus, if 0.015 in. was removed at surface "A", 0.02565 must be removed from "C" (0.015 times 1.71 = 0.02565).

VALVE GUIDES

The valves reciprocate within guides that are either integral, amounting to no more than accurately positioned holes in the head, or replaceable tubular inserts, known as thick-wall iron inserts.

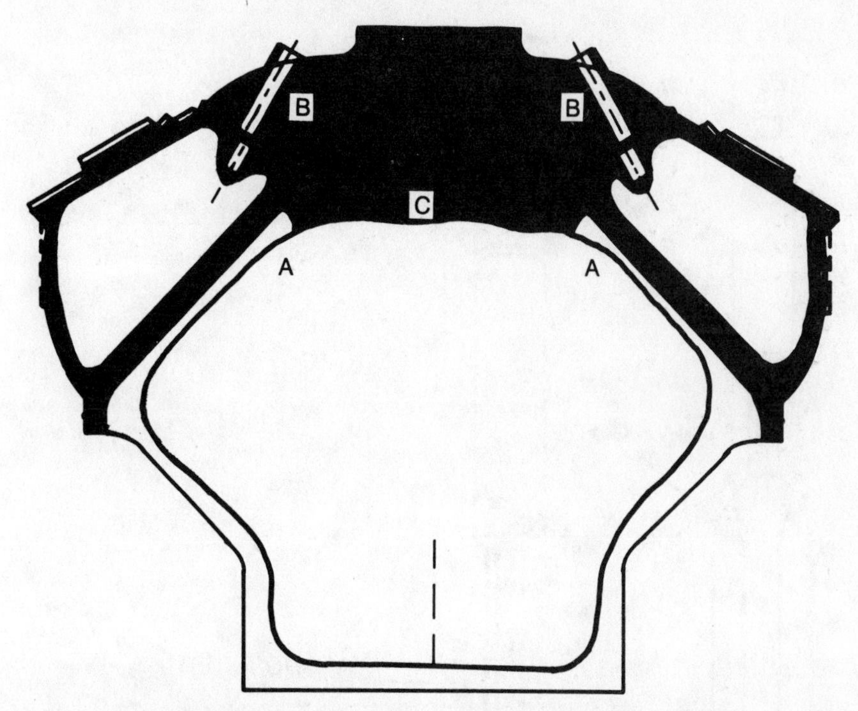

Fig. 5-9. Additional intake-manifold machining may be required if the manifold acts as the tappet cover. Failure to do this may defect the rail gaskets, causing oil leaks.

Side forces generated by the rocker arm tend to cant the valve diagonally to the guide ID, wearing away the upper and lower ends of the guide. The loss of the sharp edge at the lower end of the guide encourages carbon buildup on the valve stem and accelerates wear; the bellmouth at the upper end acts as a funnel to collect oil, which then enters the cylinder. Most shops determine the amount of bellmouth by measuring (or, if truth be told, by feeling) the amount of valve wobble. Figure 5-10 illustrates the use of a split-ball gauge. Subtracting valve-stem diameter from guide diameter gives the running clearance. Budget shops may allow as much as 0.005-in. slope; premium machine shops do not like to see much more than 0.003 in. for conventional (solid-stem) valves.

Manufacturing error or casting creep may displace the guide relative to the valve seat. The valve stem must bend to accommodate. This costs power (since some valve motion is lost as side deflection) and increases valve stem/guide wear. Valve seat grinding tools pilot on the guide, automatically restoring concentricity between the seat and guide.

Resize or Replace?

If we discount knurling, which is a quick and dirty fix, hardly appropriate for a rebuilt engine, there are five ways to deal with worn valve guides.

Oversized Valve Stems. When integral guides are worn, the machinist can ream the guide to the next oversize and fit valves with appropriately oversized stems. For example, Chrysler Slant Six valves come in 0.005-, 0.015-, and

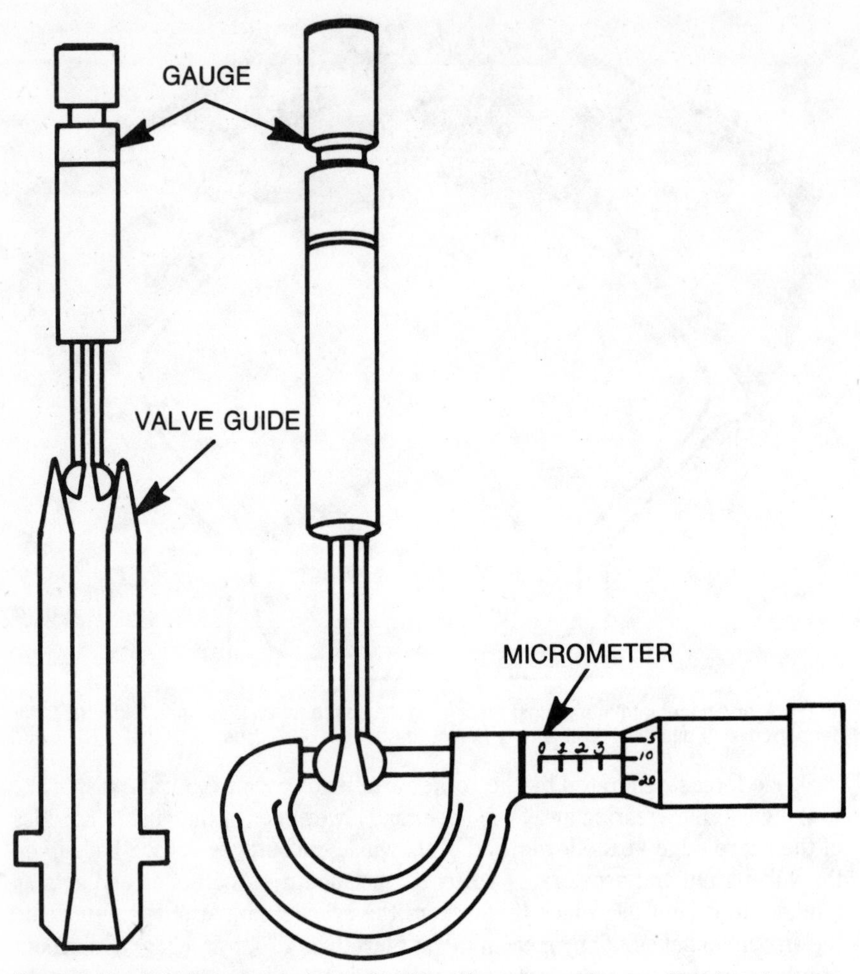

Fig. 5-10. A split-ball gauge may be used to determine valve-guide wear. Note how the guide is bellmouthed at both ends.

0.030-in. oversizes. Oversized valves are also available for some engines with replaceable guides. The fix is simple, involves minimum reworking, and maintains guide-to-valve-seat concentricity, but new valves are expensive.

Thin-Wall Inserts. The most popular repair for both integral and replaceable valve guides is to ream the guides oversize to accept thin-wall silicon or phosphor-bronze inserts, which are then finish-reamed to factory specs. The repair is inexpensive and does a fair job of maintaining seat concentricity. Cast iron inserts are also available, but, until recently, were considered inferior to silicon bronze. The advent of no-lead fuels has changed this; without the lubricity provided by lead, cast iron may be the best alternative after all.

Wire Inserts. Both Sunnen and Winona market insert repair kits that work on the Heli-Coil principle. The original insert is reamed oversize, tapped, and the insert wire is threaded into place. The guide is then reamed (Winona) or honed (Sunnen) to size. Installed correctly, wire inserts maintain original valve-seat concentricity and should be good for more than 100,000 miles.

Replaceable Guides. Some machinists have no truck with any of these methods and go back to basics with new replaceable guides. Heads that were originally equipped with integral guides are reamed oversize to accommodate the new guides—which is probably something that the factory should have done in the first place.

Replaceable guides are driven out and new ones driven in. (Aluminum heads should be heated to facilitate guide removal or, alternatively, the guides should be drilled oversize to relieve hoop tension and then knocked out.) Some inserts bottom on a flange (e.g., exhaust side, Chevrolet big-block V-8s), which determines the installed height of the insert. Others have a constant diameter and must be installed to predetermined depth. Once the insert is in place, it is reamed to size. New inserts rarely center on the valve seats, and seats must be ground to accommodate.

Clearance

Guide-to-stem clearance ranges between 0.001 and 0.003 in. for intake valves; exhaust valves are set up tighter with 0.0015 to 0.003 in. Sodium-filled exhaust valves, used in trucks and high-performance cars, like to run at about 0.004 in.

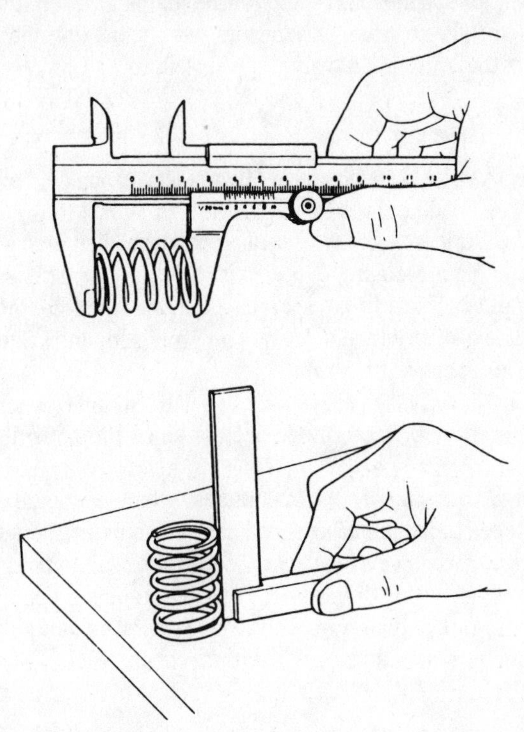

Fig. 5-11. Valve springs should be measured and should stand tall with no more than 2.3 degrees of tilt. (Courtesy Perfect Circle)

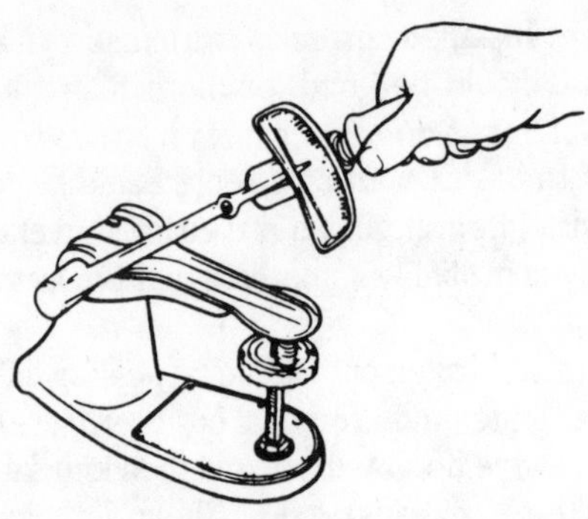

Fig. 5-12. A machinist's rule is that valve springs should develop at least 100 lb. of tension when compressed 0.3 in. for street engines.

VALVE SPRINGS

The machinist should examine the springs for pitting, flaking, and flattened ends. Springs should stand vertically—more than 2.3 degrees of tilt is an indicator of fatigue—and should be within 10 percent of the height of new springs (Fig. 5-11). As a rule, the springs are okay if they stand straight and equally tall. The acid test is to determine spring tension with the tester shown in Fig. 5-12. For street applications, a spring is good if it develops at least 100 lb. of tension when compressed 0.30 in. Some machinists shim weak springs; others play it safe and replace the springs. A valve spring failure can lunch the engine.

VALVES

Figure 5-13 illustrates the nomenclature. High-performance engines may be fitted with special valves, distinguished by materials or finish.

Sodium-filled exhaust valves—identified by large-diameter stems—use liquid sodium as a cooling medium. No special service operations are required. However, it should be noted that sodium explodes upon contact with water: the integrity of the valve stem must not be compromised, and failed valves should be discarded in an approved manner.

Stellite valves—hard-faced valves coated with Stellite (or similar alloys)—resist wear. These have become more popular since the advent of low and no-lead fuel.

Swirl polished valves have a precise radius, which eliminates stress risers, or scratches that lead to fatigue failure, and can have beneficial effect on gas flow.

Aluminized valves have their face coated with aluminum for corrosion protection. These valves, distinguished by their rough and cobbly appearance, should be replaced, rather than reground. Newly installed aluminized valves will leak until the engine starts and runs a few seconds.

FAILURE

Normal leakage carbons over the face of the valve, but unless other problems are found, such a valve is serviceable after regrinding.

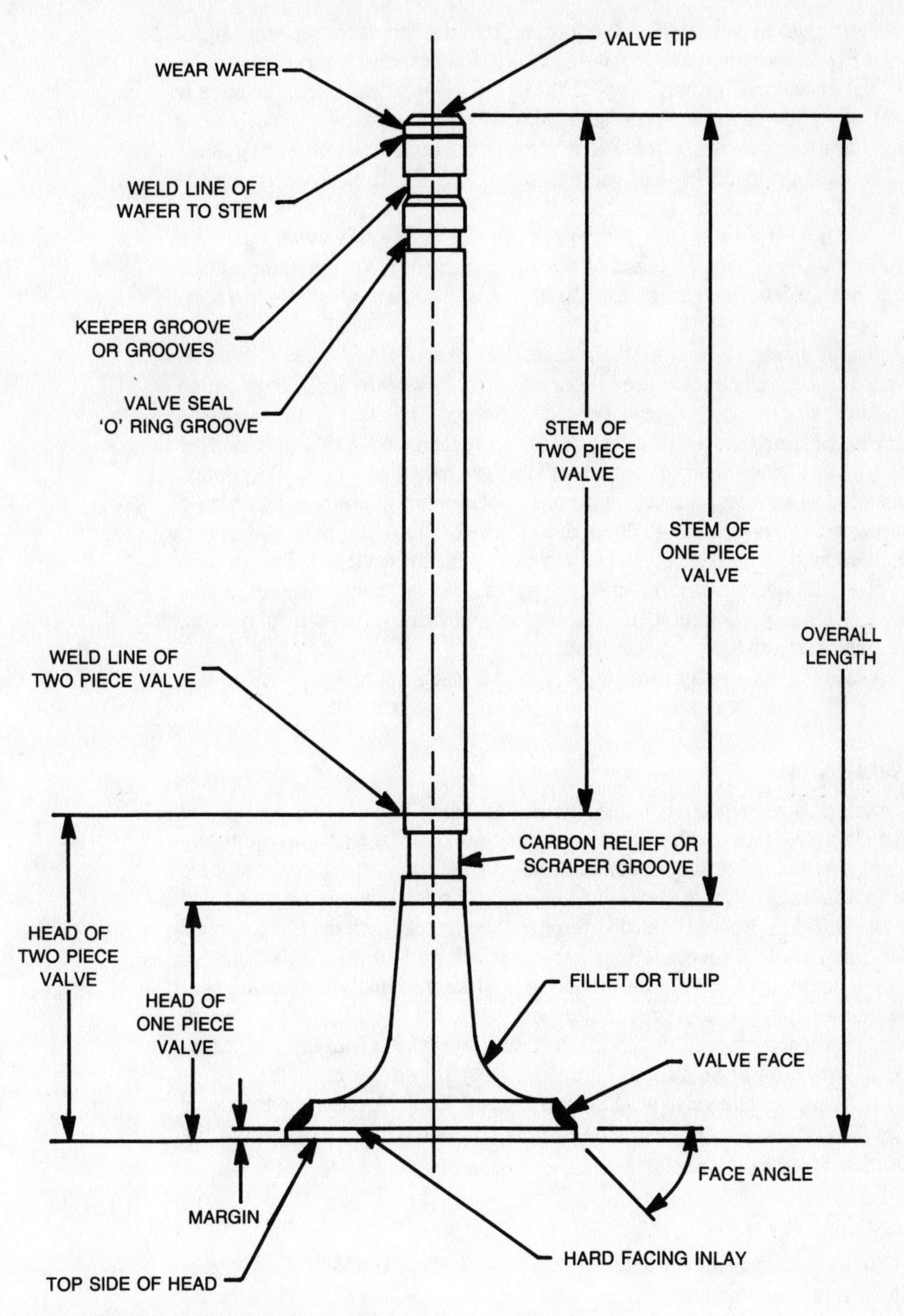

Fig. 5-13. Valve nomenclature. (Courtesy Dana Corp.)

Valve breakage usually takes the form of fatigue failure from repeated shock loads or bending forces. Fatigue failure can be recognized as a series of annular rings, not unlike tree growth rings, at the parting surfaces. Shock damage is usually caused by excessive lash (stem-tip-to-rocker clearance) or weak valve springs. Bending comes about because of nonconcentric seats or worn guides.

The most dramatic form of valve failure is the result of contact with the piston. Valve-piston collision can occur if the valve sticks (see below) or if the valve "floats" off its seat into the path of the piston. Weak valve springs are the usual cause, although excessive engine speed can have the same effect. Chronic valve float can wear the flats off the camshaft lobes, giving them a tear-drop appearance.

Valve burning, which in extreme cases takes the form of holes torched in the valve face and seat, arises because of excessively high combustion temperatures or because of improper valve cooling. The most common causes of high combustion temperatures are detonation (usually traceable to excessive spark advance), plugged or inoperative exhaust gas recirculation (EGR) systems, and manifold vacuum leaks. Insufficient valve cooling usually comes about because of excessive valve-stem-to-guide clearance. Also check for damaged seats, restrictions in the exhaust system, and scale accumulations in the cooling system.

Valve sticking, signaled by scuffs or deep scores on the valve stem, arises because of elevated valve temperatures and/or insufficient valve stem lubrication. Valve guides are the primary suspect.

Discuss these and other problems with the machinist and take corrective action.

REWORKING

Reworking a valve involves two operations: stemming, or squaring the tip, and grinding the valve face. Both of these operations are done by machine. Stemming provides a smooth, flat surface for the rocker arms to react against, and thus reduces side loads on valve stems and guides. As a rule of thumb, you can grind valve tips 0.015 in. without breaking through the surface hardening. Exceptions to this rule include Chevrolet Vega valves, which go soft after 0.005 in.

Most automotive valves are ground at a 44-degree angle for mating with 45-degree seats. The 1-degree difference is the interference angle, which ensures high seating pressures and which flattens to conform with the seat during the first few hours of operation. Figure 5-14 illustrates the concept.

A correctly ground valve will have a margin of 1/32 in. A thinner edge would overheat and could wreck the engine by igniting the mixture early, before the spark plug fires.

VALVE SEATS

Most domestic manufacturers agree with Henry Ford and run the valves against integral seats, machined directly into the head (Fig. 5-15). Integral seats are cool-running, reasonably durable when flame-hardened, and, of course, inexpensive.

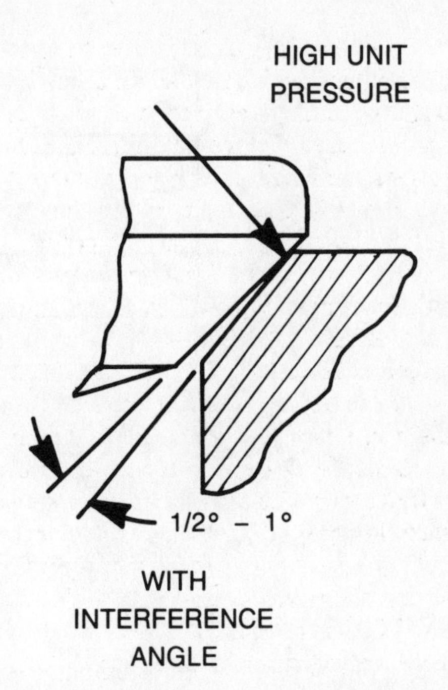

Fig. 5-14. The interference angle between newly reground valves and seats generates high unit pressure and eliminates the need for lapping. (Courtesy TRW)

Foreign engines and all engines with aluminum heads use replaceable valve seat inserts, shrunk-fit into the head. Hardened iron remains the most popular material, although stainless steel and Stellite may be specified for high-performance applications.

Fig. 5-15. Model-T block with integral valve guides (and broken head bolt).

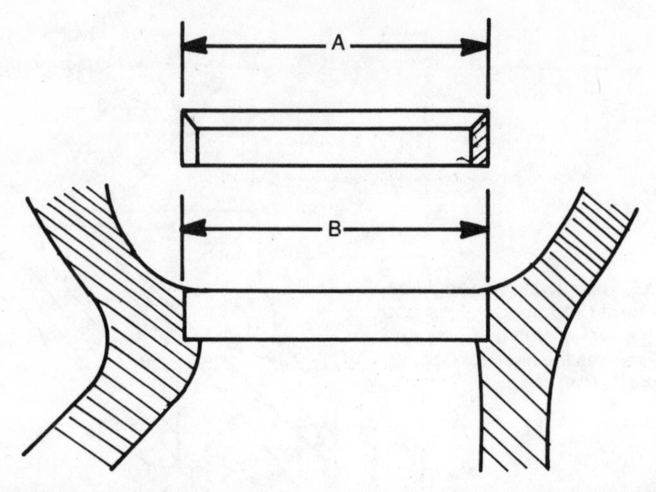

Fig. 5-16. Seat OD ("A") is always a few thousandths of an inch larger than counterbore diameter ("B") to establish an interference, or press, fit. (Courtesy Continental)

SERVICE

Heads with severely worn or burned integral seats can be counterbored for inserts. A 0.003- to 0.005-in. interference between the counterbore OD and the insert ID is sufficient to keep the cast-iron inserts in place (Fig. 5-16). Stainless steel and Stellite inserts expand less with heat than the head and, in in addition to an interference fit, require peening.

The machinist uses a puller to remove existing inserts (Fig. 5-17). Inserts in cast-iron heads lift easily; aluminum heads may have to be partially cut away before they can be pulled. It also helps to heat the head. Counterbores should be reamed to the next oversize for an interference fit of 0.003 to 0.005 in. for iron and 0.006 to 0.008 in. for aluminum. Shrink-cooling inserts with dry ice or Freon makes installation easier in cast-iron heads; and is mandatory for aluminum (Fig. 5-18).

In the absence of factory specs to the contrary, the seat should be ground at 45 degrees (with the 1-degree interference on the valve side) to a width of 1/16 in. wide on intake valves and 3/32 in. on the exhaust valves (Fig. 5-19). Narrower seats pound out and wider seats fail because of accumulated carbon. In addition, valves should overhang the seats by about 1/64 in. (Fig. 5-20). At operating temperature, the valve elongates, moving the contact area down the face, toward the stem. Valves that seat low on the face may fail to close.

These conditions are met with a three-angle seat grind (Fig. 5-21). The first cut establishes the 45-degree seat angle; the second, or topping, cut defines the upper edge of the seat; the 60-degree throating cut fixes the lower edge of the seat. Most machinists mark the valve with Prussian Blue to determine where the seat contacts the valve face. Topping lowers the contact area; throating raises it. Of course, seat width must be kept within specification.

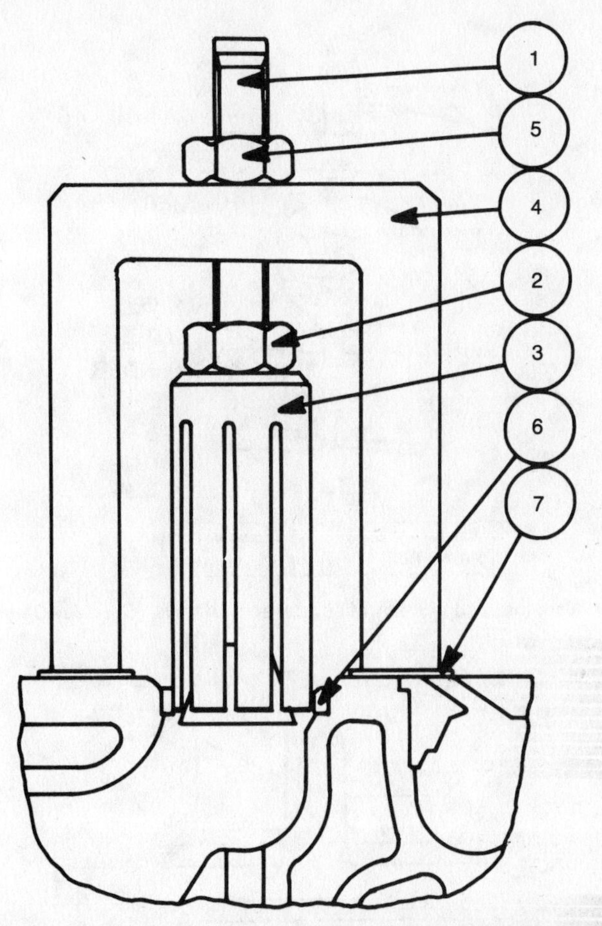

Fig. 5-17. A valve seat insert puller. As the nut (5) is tightened, the shaft (1) draws the cone up and spreads the claw (3) against the insert (6). Copper pads (7) protect the head. (Courtesy Chrysler-Nissan)

In years past, no valve job was complete unless the valves and seats were lapped together with grinding compound. This practice is no longer recommended.

ASSEMBLY

The machinist will assemble the valves, springs, keepers, and seals. One critical measurement remains—valve-spring assembled height (Fig. 5-22). Two keepers and a retainer are mounted on each valve, and the distance between the underside of the retainer and the spring seat is measured. Refacing typically raises the valve in the head by about 0.030 in. and relaxes spring tension by a like amount. A spring shim installed under the spring restores tension (Fig. 5-23).

There is a special problem associated with ohv engines equipped with hydraulic lifters that valve-spring assembled height only indirectly addresses. The

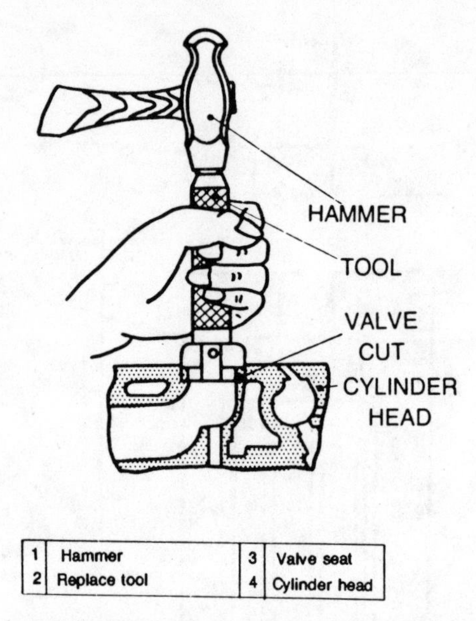

Fig. 5-18. Inserts are installed with any of several special tools. (Courtesy Chrysler-Nissan)

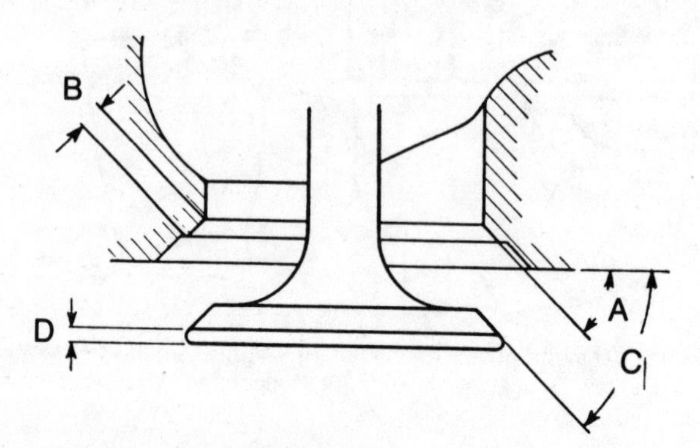

Fig. 5-19. A good valve job should look something like this: A = 45 degrees, B = 1⁄16 in. on intake, 3⁄32 in. on exhaust, C = 44 degrees, D = 1⁄32 to 1⁄16 in.

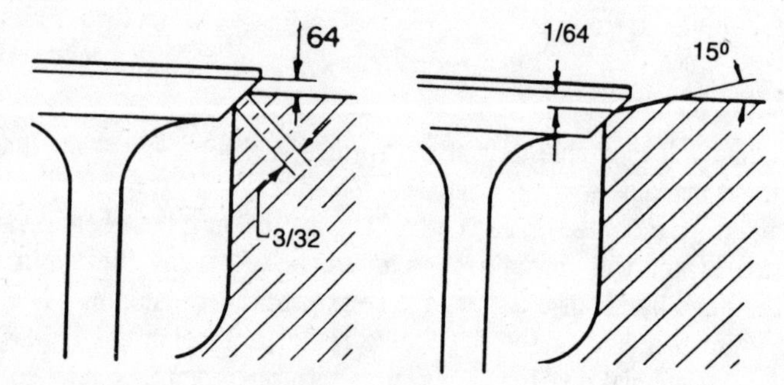

Fig. 5-20. The valve should overhang the seat by about 1⁄64 in. (Courtesy TRW)

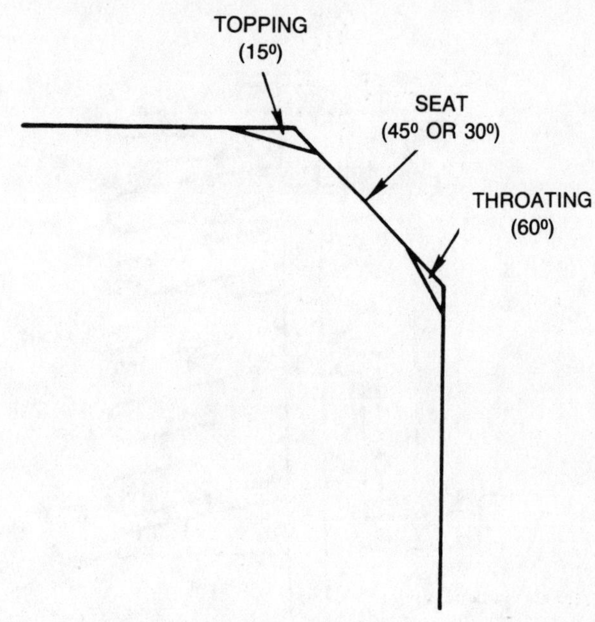

Fig. 5-21. The depth of the topping cut determines the amount of valve overhang. (Courtesy Dana Corp.)

dimension of concern is assembled valve-tip height as measured from the head gasket surface.

Resurfacing the head or block forces the pushrod deeper into the hydraulic lifter; grinding the valves or seats has the same effect.

The lifter can compensate, but only over a narrow range. The average lifter plunger has a stroke of 0.150 in. If the plunger is centered at mid-travel in its bore, it can move 0.075 in. before it bottoms and holds the valve off its seat. But lifter plungers are rarely centered, and available stroke may be less than indicated. A deep valve and seat grind can take out 0.100 in.

Excessive valve-tip height puts the rocker arm at a bad angle, causing it to skew the valve to the side. Guide wear increases dramatically.

The assembled spring height measurement is helpful to the extent that it shows how much metal has been removed from the valve face and seat. But it does not indicate how much metal has been lost as a result of cylinder head/block resurfacing and valve stemming.

A good machinist will avoid the problem by limiting cuts on the gasket surfaces and replacing worn seats, rather than grinding them in. He has a very limited choice of ways to correct excessive valve-tip height: no more than 0.015 in. can safely be removed from the tip, and a new seat can be installed to lower the valve and thus raise the plunger. As a last (and expensive) resort, head and block thickness can be restored by spray welding.

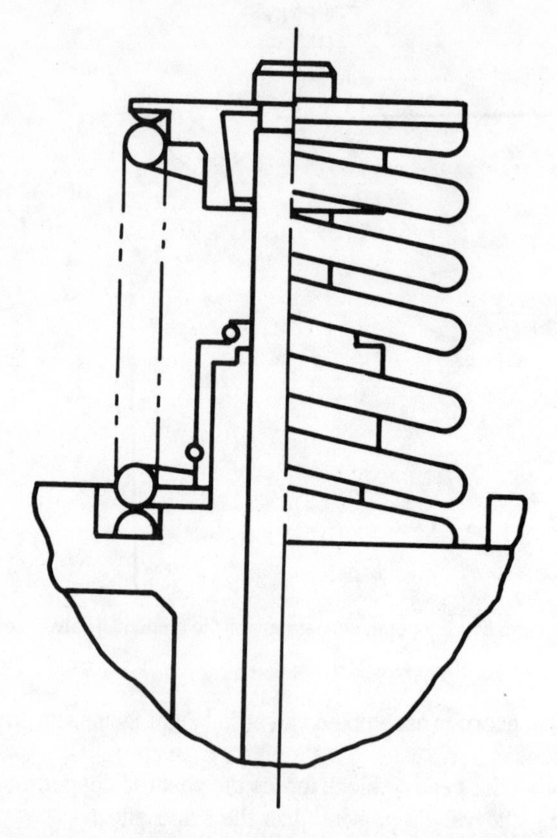

Fig. 5-22. Valve-spring installed height is the distance between the spring and the underside of the spring retainer. Often the factory neglects to provide this specification and spring height must be determined by measurement during disassembly. (Courtesy Chrysler-Nissan)

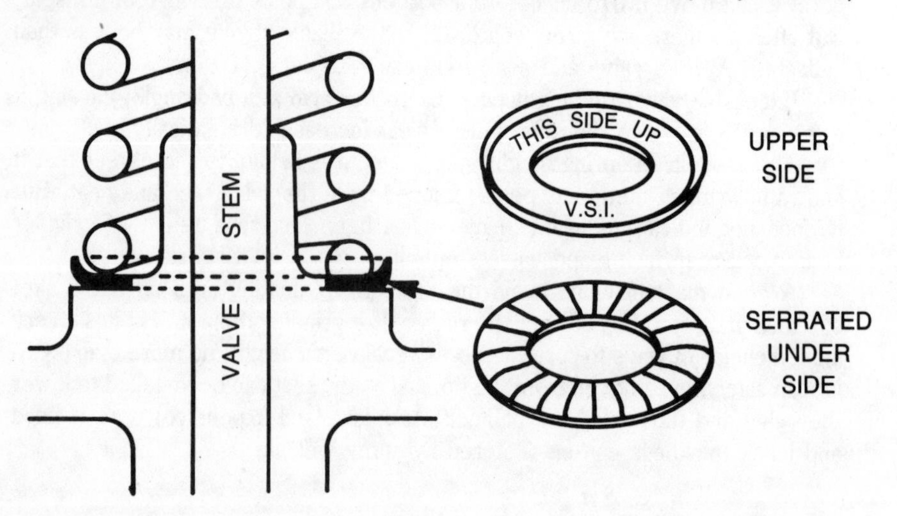

Fig. 5-23. A typical valve and seat grind adds about 0.030 in. to valve-spring installed height. An appropriately sized shim will restore spring tension. (Courtesy Dana Corp.)

SEALS

Two styles of valve stem seals are used. Most engines employ umbrella, or deflector, seals as shown in Fig. 5-24A. Some manufacturers, such as GM and Toyota, back up the umbrella seal with an O-ring just under the valve keepers. Positive seals, which seat against the valve guide boss, give improved oil control (Fig. 5-24B). These seals can often be retrofitted to engines that were originally equipped with umbrella seals. But consult your machinist before changing over to positive seals: street engines may not benefit from this sophistication, and there have been reports that these seals are too positive in some applications and starve the valves for oil.

NO-LEAD FUEL

Most pre-1975 auto engines (and industrial engines built into the early 1980s) were designed to burn gasoline that contained as much as 4 or 5 grams of tetraethyl lead (TEL) per gallon. Lead was both an inexpensive octane booster and a source of lubrication for the exhaust valve. It was also a health hazard. Consequently, the TEL content in leaded gasoline has been steadily reduced and currently stands at 0.1 gram per gallon. Exxon, the nation's largest refiner, devotes it entire production to unleaded gasoline. Other refiners are expected to follow Exxon's example.

There are basically three problems associated with no-lead or low-lead gasoline. One is the lower effective octane rating. High-performance engines, built in the 60s and early 70s, must be detuned to run on modern fuel. The best way to do this is to lower the compression ratio, either by using late model, big-chamber heads or by changing pistons. A compression ratio of about 8.5 to 1 is about all that can be safely tolerated, although one can go a point higher by using water/methanol injection and spark retardation.

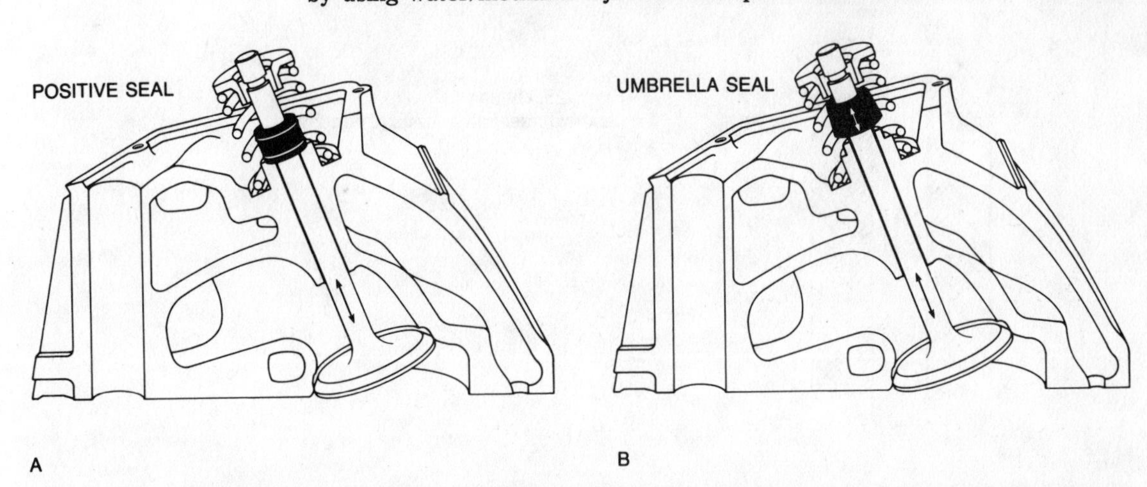

Fig. 5-24. Umbrella seals move up and down with the valve stem, deflecting oil away from the guide ("A"). Positive seals mount on the valve guide boss, wiping the stem as it reciprocates ("B"). As mentioned in the text, positive seals can be fitted to most domestic engines with a minor cleanup operation on the valve guide boss OD. Currently the best available material for seals of either type is Viton, which withstands temperatures of up to 450 degrees F. (Courtesy Fel-Pro Inc.)

Both old and new engines are affected by the tendency of no-lead gasoline to increase carbon deposits. While driving style has a lot to do with it, no-lead fuel deposits start to build after 10,000 miles and may continue until the engine self-destructs from detonation. (Carbon deposits from high-lead gasoline stabilize after 5,000 miles.)

Carbon-induced compression increases were apparently behind GM's decision to increase deck height on the 151 cross-flow engine. In 1981, the deck height—the distance from the centerline of the crankshaft to the top of the block—was 9.140 in. By 1982 deck height grew to 9.200 in., thus lowering the compression ratio. GM and aftermarket suppliers furnish a 20-in. thicker head gasket for 1980-81 engines.

The major problem associated with low or no-lead fuel in older engines in valve seat recession, illustrated in Fig. 5-25. Each time the valve closes, tiny welds form between the valve and seat. When the valve opens, the welds stretch and break, leaving jagged deposits on the valve face. These deposits abrade the seat and encourage additional welding and more rapid recession. Metal lost to the seat collects on the lower valve stem and distorts the guide.

Valve rotators, sometimes fitted to these older engines, only make matters worse. Heavily loaded engines can fail within a few thousand miles.

Any engine (regardless of date of manufacture) that uses no-lead fuel demands accurate seat-to-guide alignment with concentricity held to 0.001 in. Wide seats also help by reducing valve temperatures.

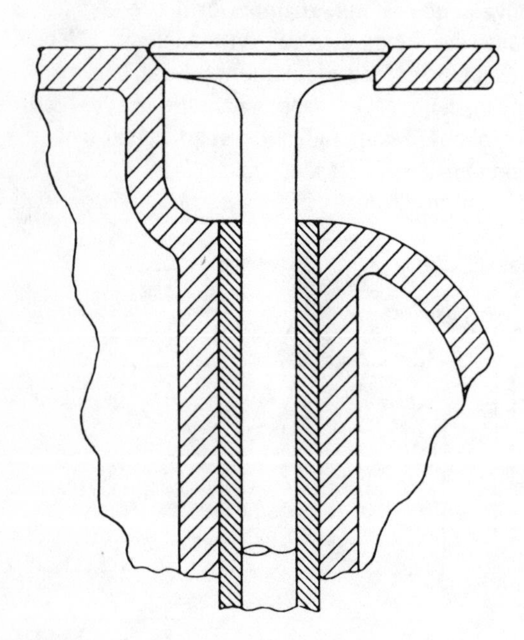

Fig. 5-25. Unleaded fuel tends to recess exhaust-valve seats. (Courtesy TRW)

Older engines with integral or soft-iron valve seats should be updated with Stellite inserts and high nickle content valves, at least on the exhaust side. The conversion costs about $200.

As an interim measure, TEL can be purchased at speed shops and mixed with the fuel. Other, presumably safer, additives, such as Unocal 76 Valve Saver, are also available.

6
CHAPTER

Teardown and Rework

MOST PREMIUM MACHINE SHOPS LIMIT THEIR ACTIVITY TO ACTUAL MACHINE operations; the customer is responsible for teardown and final assembly. Typically, the customer delivers the bare block, together with pistons, rods, rod caps, crankshaft, flywheel, and main bearing caps, to the shop. The machinist hot-tanks the iron parts, then determines cylinder-bore and crankshaft-journal wear, and makes an estimate of charges.

Logistics are simplified somewhat if replacement parts are purchased through the shop. Machine work usually requires about a week to complete.

The block must be dismantled completely, so that all that remains are the camshaft bearings and core plugs (Fig. 6-1). Work deliberately, examining each part as it comes off for insight into engine condition. A cracked ac compressor bracket can explain a knock at idle; an external oil leak will alert you to a problem that may require more than a new gasket to correct; an uneven wear pattern on the main bearing inserts can mean main bearing misalignment or a bent crank; and so on. Save the old parts, including the old gaskets, for comparison with the replacement.

In the midst of all this activity, take time to become familiar with the oiling system. Drawings in the service literature help, but are no substitute for physically tracing the circuits and noting the presence of jiggle pins (Ford. 390), restrictors (almost universal), and air bleeds (e.g., the ported oil gallery plugs on the front of Chevrolet big-block V-8s—if these plugs are installed at the rear, massive hemorrhaging results).

STANDARDS

Auto engines were once built to agreed-upon standards. There were few running changes during the course of a model year; parts for a given engine were

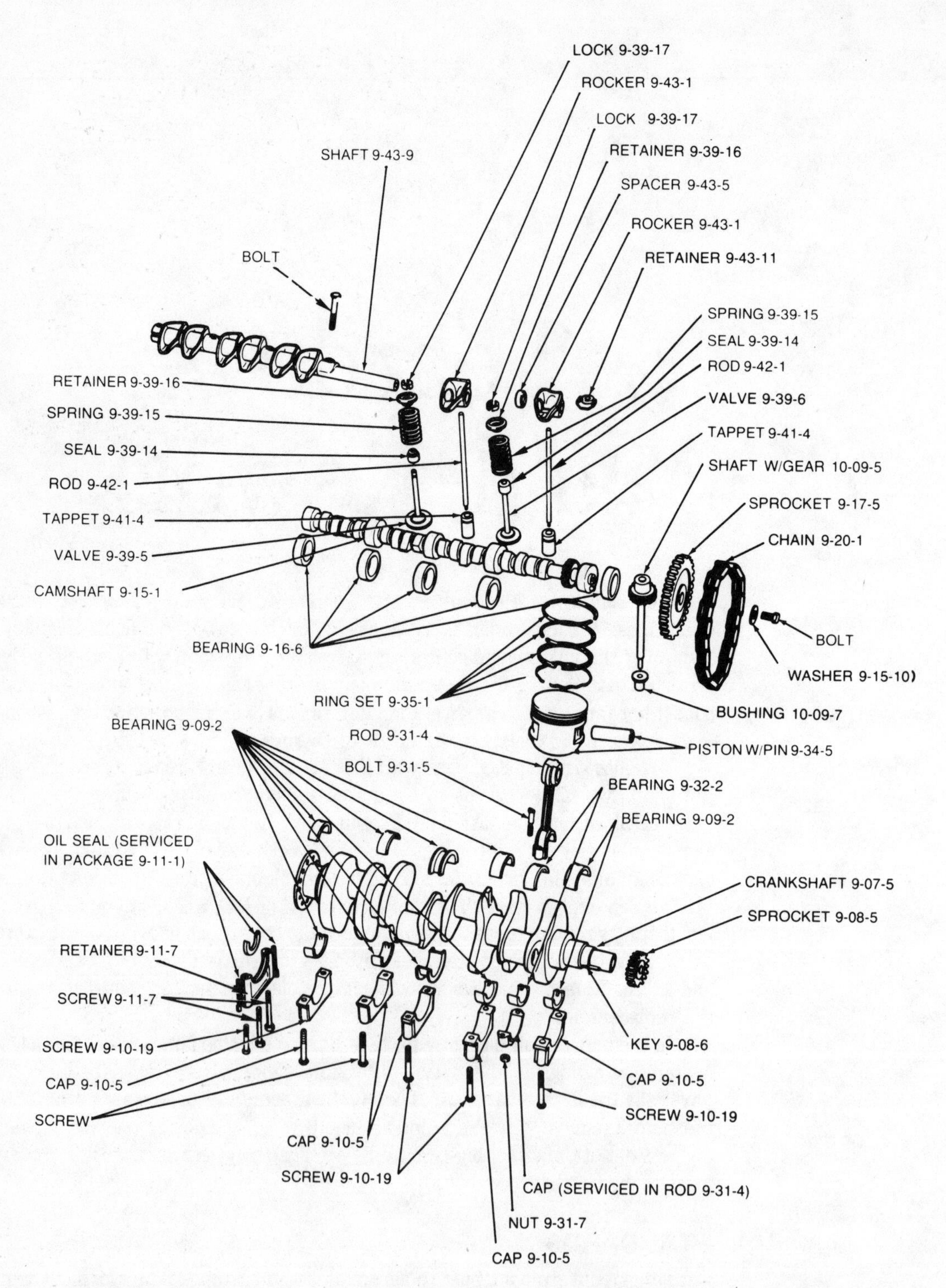

Fig. 6-1A. Nomenclature, internal parts-Chrysler small block (273, 318, 340, 360).

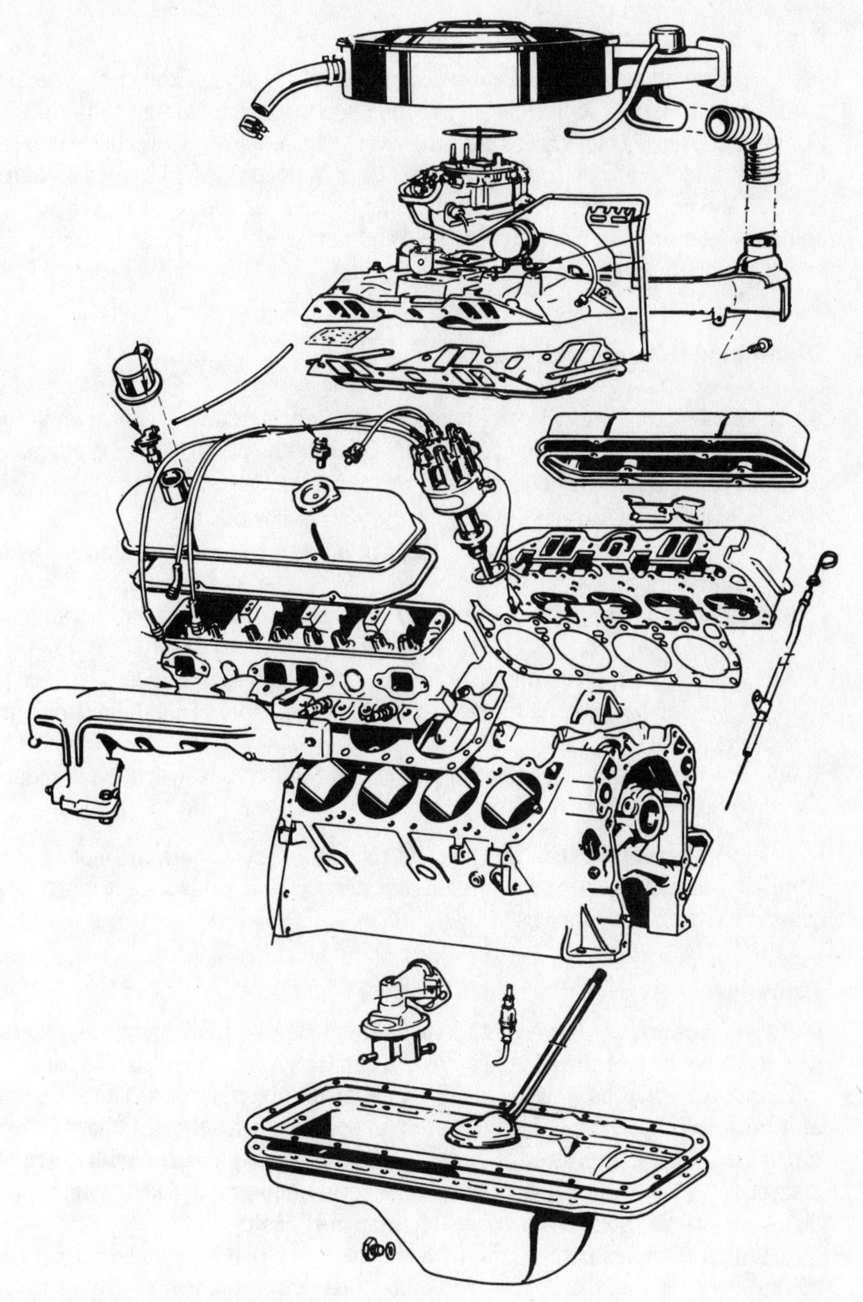

Fig. 6-1B. External parts arrangement, typical of ohv V-8 engines.

almost always dimensionally interchangeable with other engines of the same type; and SAE (Society of Automotive Engineers) fasteners were the norm. All that has changed, and the rebuilder's life is now far more complex.

Production Changes

Many engines undergo modification during the model year. Factories generally notify their dealers of such changes, but the word is slow to get out to the trade. Some of the least stable areas for American engines are timing chain covers, head bolts, and crankshaft oil seals. As a case in point, GM 112 and 121 fours and 173 V-6 engines were originally equipped with fiber rope rear oil seals. The company later went to a split rubber seal and now uses a one-piece (radial) seal, which should be retrofitted to older models.

Dimensional Abnormalities

As an economy measure, Detroit scraps fewer engine components than in the past. Machining errors are "corrected" by finishing the part to the next oversize. Sealed Power Corp. has documented instances of the following:

- ☐ Pushrods of varying lengths in the same block
- ☐ One or more valve lifter (tappet) bores oversized and fitted with oversized lifters
- ☐ All main bearing bores oversized and special, thick-walled bearings installed
- ☐ One or more rod and main bearings with undersized IDs
- ☐ Cylinder bores honed slightly oversized and special, high-limit pistons installed
- ☐ A few cylinder blocks with deck heights 0.030 in. lower than standard (nonstandard head gaskets return compression ratio to normal)

Some aberrant engines are marked to that effect; others are not, and it is up to you and your machinist to make certain that replacement parts are compatible with the originals.

Fasteners

Many engines are assembled with a mix of inch-standard and metric fasteners, which are not interchangeable. Figure 6-2 illustrates the nomenclature for both systems. Note that bolts are coded for yield strength, which is a critical parameter for many applications. Metric fasteners sold through auto parts houses are, for the most part, unmarked (Grade 1 or 2) and should be used with extreme discretion. It should also be noted that, grade-for-grade and diameter-for-diameter, metric bolts twist off easier than SAE bolts.

A further complexity arises with metric bolts that, depending upon the country of origin and the time of manufacture, may conform to any of three standards. European bolts generally follow the DIN standards or the ISO modification of the DIN. Most Japanese bolts made during the last decade follow ISO, but there exists a JIS standard, more or less uniquely Japanese. You may run into trouble replacing a JIS head bolt with a large-headed ISO number. When these difficulties arise, forget the aftermarket and go to a dealer or junkyard.

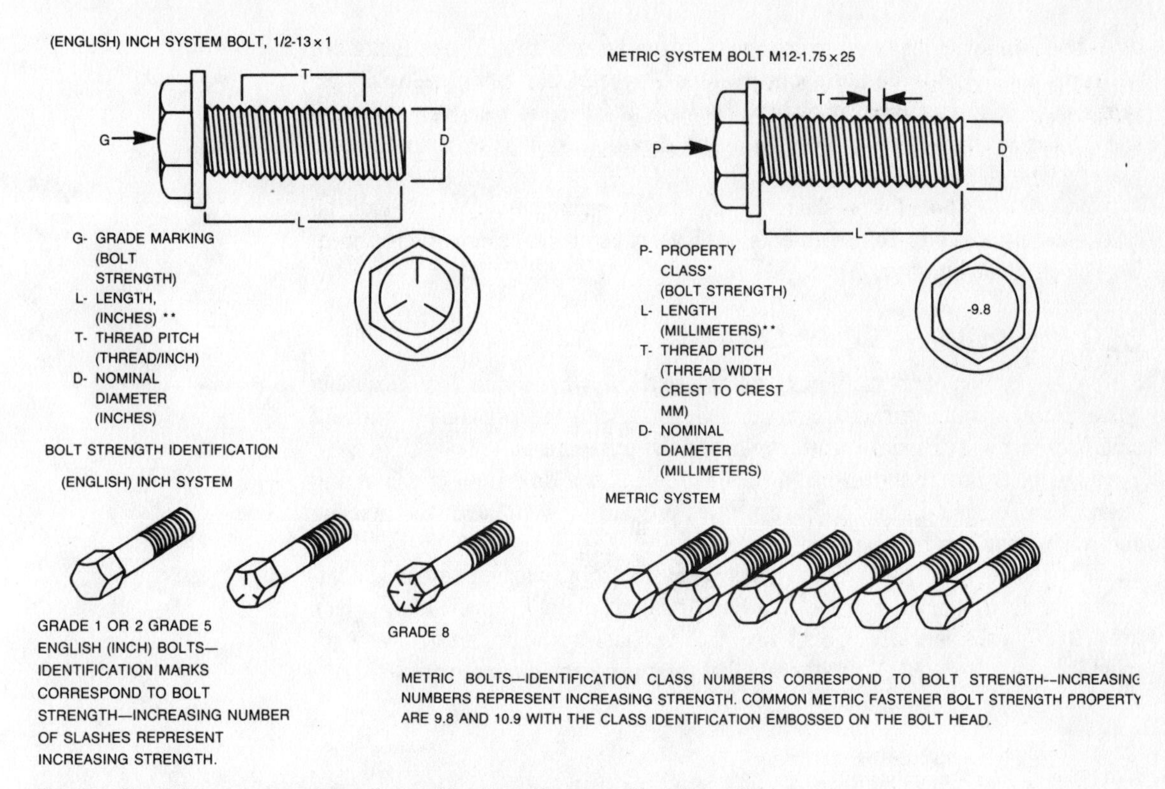

Fig. 6-2. English inch and metric bolt identification. (Courtesy Ford Motor Co.)

BLOCK INSPECTION

Inspection of the block is a continuous process, carried on as subassemblies are removed. Begin by examining the deck (head gasket interface) for cracks, leak paths across the head gasket, and indentations caused by gasket fretting. Most head gasket failures are not the fault of the gasket, loose fasteners, or a warped head. Detonation is the culprit.

An engine that has been run for thousands of miles with a blown head gasket can develop notches between adjacent cylinders, which can only be repaired by welding. You can measure deck trueness with a precision straightedge and feeler gauges, as described in the previous chapter.

Using the flywheel for leverage, turn the crankshaft and inspect the cylinder bores for scratches, localized scoring, and pitting. Run your fingers over the ridge at the top of the bores, which defines the upper limit of ring travel. The ridge (less carbon accumulations) represents the original cylinder diameter; the undercut below the ridge is roughly equivalent to amount of wear on the bore. Some ridge is normal, but a fingernail-hanging precipice usually means that the cylinders must be rebored and fitted with oversized pistons. Of course, the machinist will make the final determination.

Carefully examine the valley area on V-type engines. GM 181 blocks sometimes leak coolant through cracks above the lifter bores; Ford 400M blocks have been known to leak oil from cracks in the lifter bores. These faults are not repairable. And to be fair about it, Rolls-Royce blocks can develop cracks at the base of the cylinder sleeves. This type of failure is relatively common and, so far as the author knowns, remains mysterious. It is as if some force were pulling the engine apart vertically.

Other likely spots for cracks are at the main bearing webs. Stress-induced cracks that involve structural elements, such as webs or ribs, cannot be repaired. The block must be replaced.

CLUTCH/PRESSURE PLATE

Scribe the position of pressure plate relative to the flywheel edge as an assembly guide. Some manufacturers balance the flywheel and the pressure plate as a unit. Working in a crisscross pattern, loosen the pressure-plate bolts (Fig. 6-3). Turn the bolts out no more than a turn and a half at a time until clutch spring tension is released. Failure to observe this precaution will distort the housing and might result in harsh clutch action.

WARNING—Nearly all clutch disks are made from an asbestos-based composite. While it is argued that clutch (and, for that matter, brake) dust is less lethal than fibrous asbestos, OSHA says that mechanics who undertake this work must wear an approved respirator and that their asbestos-contaminated overalls

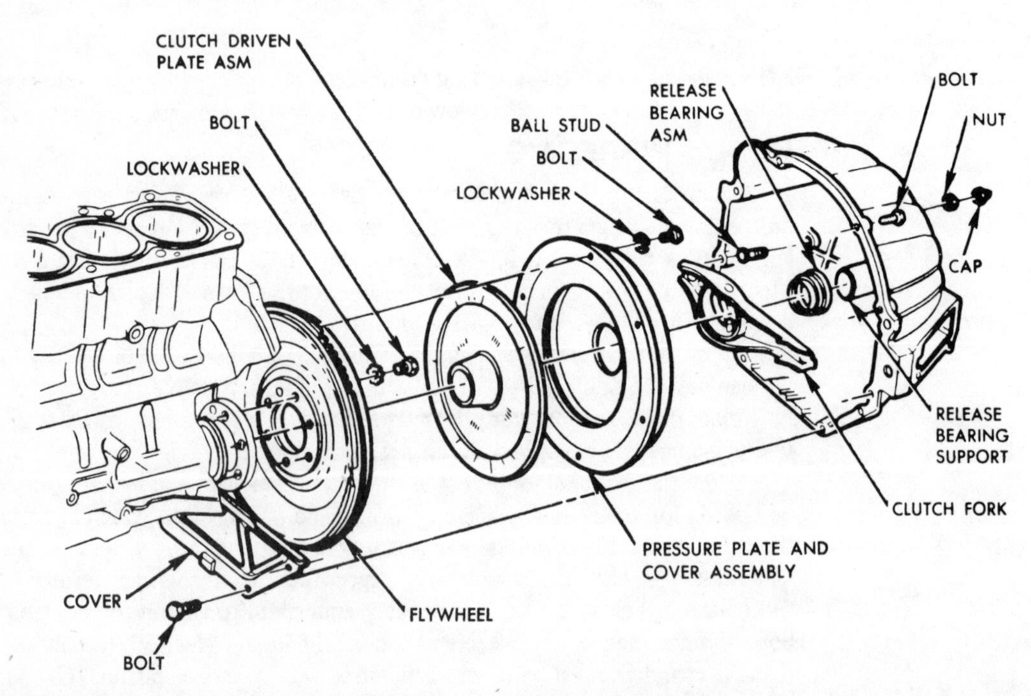

Fig. 6-3. Clutch, pressure plate, and bellhousing. Note that the clutch disk may have a front and rear face, defined by the distance the splined hub extends out of the unit. Also of interest in this drawing is the open deck construction of the engine block. Water jackets are roofed by the cylinder head. The new GM Quad-4 is built on the same pattern. (Courtesy Chevrolet Motor Div., General Motors Corp.)

should be washed separately from other clothes. The washing machine then must be cycled several times while empty.

Limit dust exposure by hosing down the clutch before removing the pressure plate and after disassembly. The runoff must be caught in a drain.

The clutch disk, throw-out bearing, and pilot shaft bearing are replaced as a matter of course. Too much labor is involved to risk trusting these parts. If, for some reason, the throw-out bearing is to be reused, merely wipe its external surfaces with a shop rag. The bearing is packed with grease and should not be exposed to solvent.

The pilot shaft bearing is pressed into the flywheel hub, where it supports the free end of the transmission input shaft. Foreign engines are often fitted with a needle bearing; Detroit prefers bushings. Either type can be extracted with a bearing puller or forced out by hydraulic pressure. Pack the bearing cavity with chassis grease and start a pin, sized to bearing ID, into the bearing. (A scrap transmission input shaft or a clutch alignment tool also can be used.) A sharp hammer blow on the end of the pin will pressurize the grease and force the bearing out of its cavity (Fig. 6-4).

The pressure plate must be replaced if warped, deeply scored, discolored, or heat-checked, with shallow, radial cracks running out from the center of the plate. Even when these defects are not present, a new or professionally rebuilt pressure plate will pay dividends in smoother clutch action.

OIL PAN

Loosen the oil pan bolts, working diagonally from the center out. The gasket seal can be broken by a sharp blow on the side of the pan with a rubber mallet.

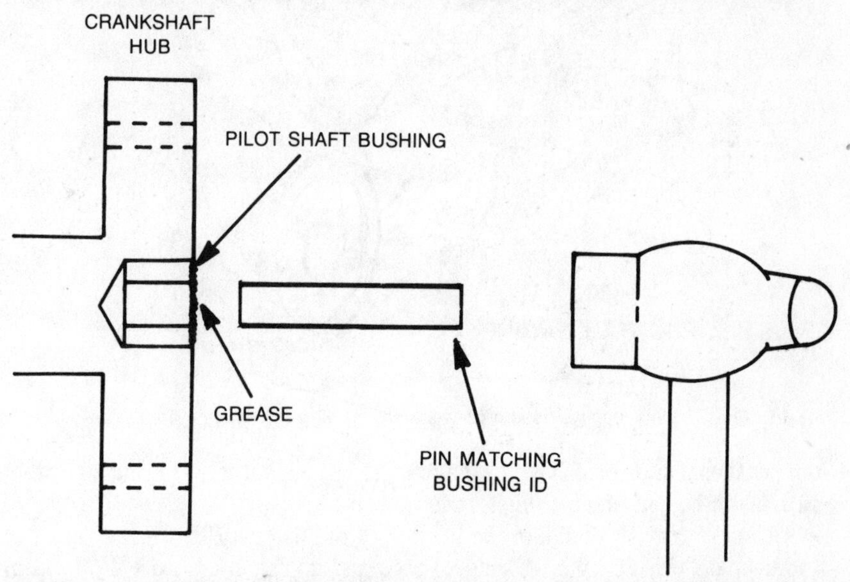

Fig. 6-4. The pilot shaft bushing can be extracted hydraulically. Use heavy grease as the working fluid and a length of shafting sized to bearing ID as the ram. Use wads of grease-impregnated paper on needle bearings.

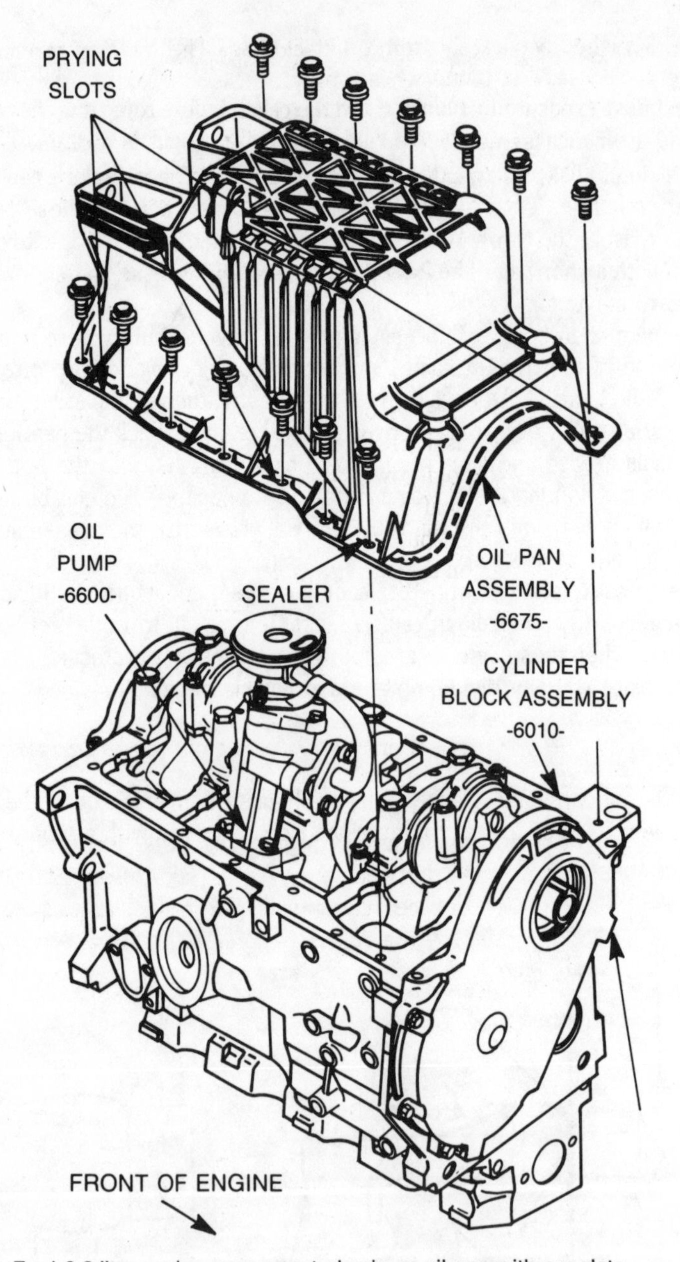

Fig. 6-5. Ford 2.3-liter engine uses a cast aluminum oil pan with pry slots.

Some cast aluminum pans have raised pry slots at the corners. Prying on a steel pan is almost guaranteed to distort it (Fig. 6-5).

Strata of grayish sludge in the bottom of the pan usually are the result of condensation, which can be traced to stop-and-go driving or to PCV system failure. Vacuum leaks or a faulty PCV valve can block air circulation through the crankcase, allowing condensation to collect. Crankcase air leaks also can defeat the system. Standing water or a mayonnaise-like emulsion of oil and water are more serious symptoms and point to a cooling system leak. Check for the presence of oil in the radiator and have the block pressure tested.

Oil and sludge can be removed with Gunk and kerosene; external road grime responds to hot water and Tide. Remove all traces of the gasket from both the pan and block rails. 1980-81 Cadillacs and a few other vehicles used plastic oil pans, gasketed with RTV (room temperature vulcanizing) silicone. Carefully scrape the sealant from these pans with a sharp knife. Do not use a wire brush or wire wheel on plastic.

FLYWHEEL/FLEXPLATE

Make assembly reference marks on the flywheel (manual transmission) or flexplate (automatic transmission). Working in the familiar crisscross pattern, progressively loosen the hold-down bolts. A short segment of two-by-four wedged between the crankshaft and block will prevent the crank from turning.

A thoroughgoing engine rebuild includes resurfacing the flywheel, preferably with a grinder that centers on the crankshaft flange. Flywheel lathes or modified brake-drum lathes chatter at the rough spots and cannot handle cup-type 'wheels.

Typical flywheel problems include the following:

- ☐ Wear. Evidenced by grooves in the flywheel face and matching grooves in the disk.
- ☐ Heat Checks. Superficial cracks, often running radially across the flywheel face.
- ☐ Discoloration. Local hard spots caused by clutch slippage. The frictional surface overheats rapidly as the surrounding metal draws off the heat.
- ☐ Runout. Misalignment between the flywheel and crankshaft hub. Have your machinist determine runout with the 'wheel assembled to the crankshaft (Fig. 6-6). As a rule of thumb, 0.004 in. is the maximum

Fig. 6-6. Runout should be checked with the flywheel torqued to the crankshaft hub.

allowable; 0.005 in. will cause clutch chatter in most applications. Any grinder that references to the crankshaft hub will correct the problem.

- ☐ Dish. Forged steel flywheels—almost universal on modern cars—tend to assume a concave profile, as the result of heat developed in normal operation. This increases the effective width of the flywheel and, when coupled with a new clutch disk, makes disengagement problematic. Dish can be detected with a straightedge and a feeler gauge.

In addition, examine the bolt holes in the flywheel or flexplate for cracks, which often go hand-in-hand with transmission mount failure. *Discard any unit that is cracked.*

Pinion/Ring Gear

Flexplates have integral ring gears; most flywheels employ a discrete ring gear, shrunk-fit to the wheel. Shrunk-fit gears can be replaced, but unless you have access to an oxyacetylene torch, it is usually more economic to purchase a good used flywheel from a wrecking yard. New pinion gears are available and, unlike every other gear train in the engine, can be successfully run against a used mating gear. Starter bushings and brushes should also be replaced at this time.

The shaded areas in Fig. 6-7A illustrate the acceptable wear pattern. Note that ring-gear teeth retain their square tips and that starter pinion teeth are blunt. Both gears should be replaced when ring-gear teeth are rounded off and pinion teeth are sharpened (Fig. 6-7B), or when the pinion has milled the sides of the ring gear (Fig. 6-7C).

ABOUT STARTERS

While this book is primarily concerned with engine rebuilding, it is not immune to the "ankle-bone-connected-to-the-shin-bone" syndrome. There is little point in building an engine that can't be started, or that eats ring gears.

The quick and easy way to resolve starter difficulties is to purchase a "rebuilt" unit. But this can be expensive—the wholesale (jobber) price for a rebuilt GM diesel passenger car starter is about $100, and some foreign starters are even more expensive.

If the starter is electrically sound and the problem is mechanical, as evidenced by excessive pinion/ring-gear wear, the unit can be "rebuilt" in an hour or so at a parts cost of about $15.

Construction

Figure 6-8 illustrates the two engagement mechanisms—the inertially activated Bendix and the electromagnetic solenoid. The Bendix, with its fatigue-prone spring and stops, is replaced as a unit. The solenoid gear can be purchased as a separate part.

Chrysler geared starters are among the most complex and most expensive to rebuild, but parts are readily available (Fig. 6-9). As when approaching any

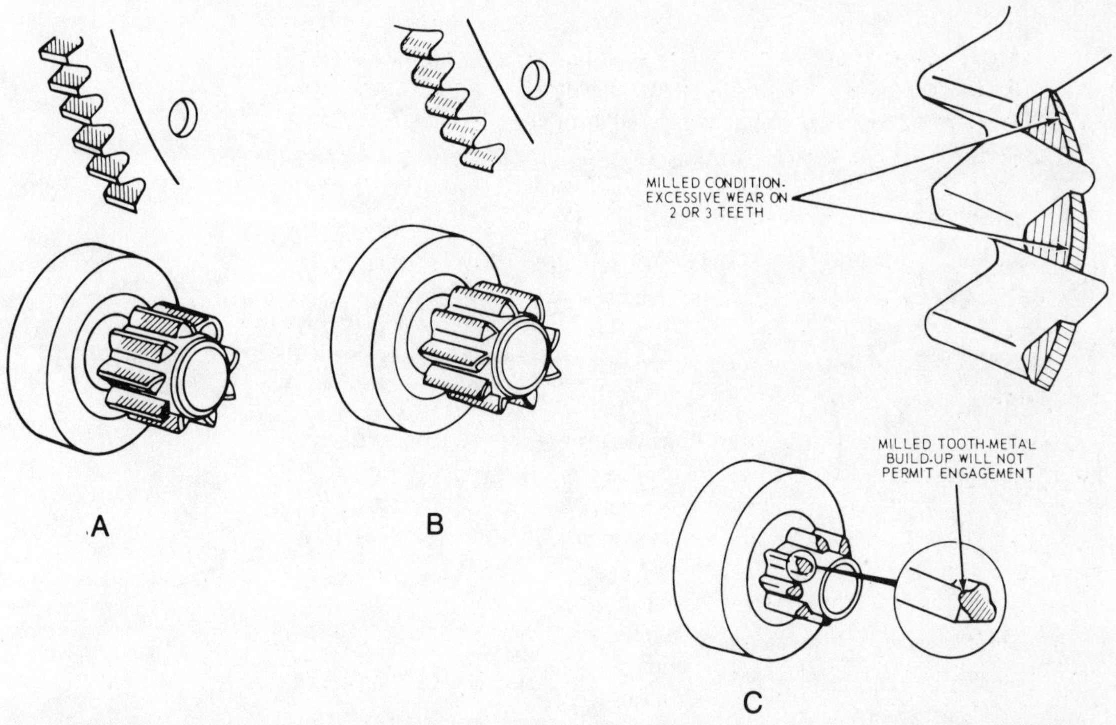

Fig. 6-7. Starter pinion/ring gear wear patterns. Wear shown at A is normal, as indicated by squared-off ring gear teeth and rounded pinion teeth. These gears may be reused. B and C illustrate gears that should be replaced, either because ring gear teeth are rounded and pinion teeth have worn to knife edges or because of severe end milling that reduces the tooth width. Flywheel ring gears are renewable, although it is often as economic to purchase a used assembly. (Courtesy Ford Motor Co.)

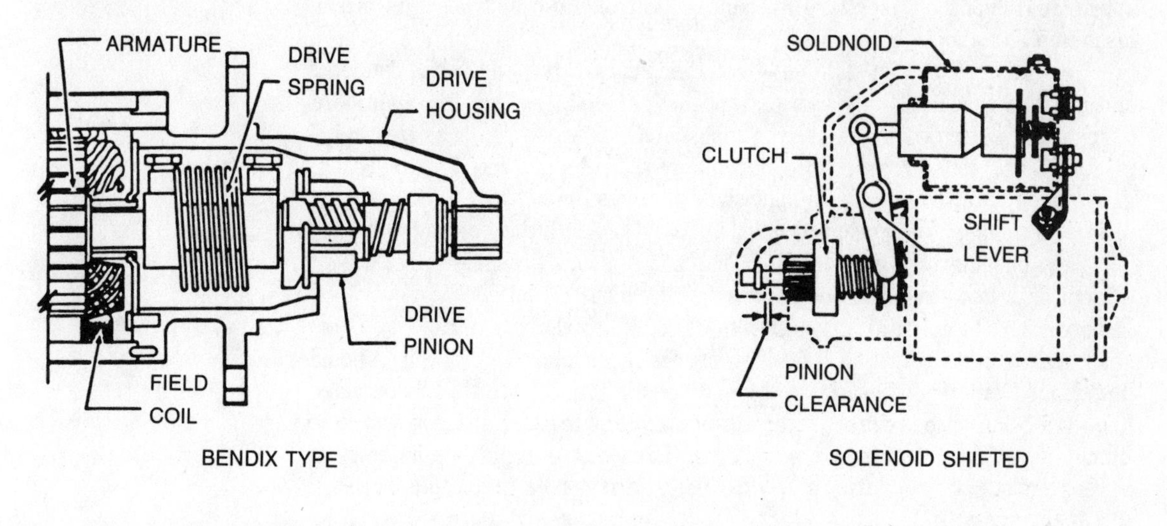

Fig. 6-8. The Bendix starter drive is inertially engaged with the ring gear; the solenoid-type is engaged by means of an electromagnet, which also closes contacts that energize the starter motor. (Courtesy Kohler of Kohler)

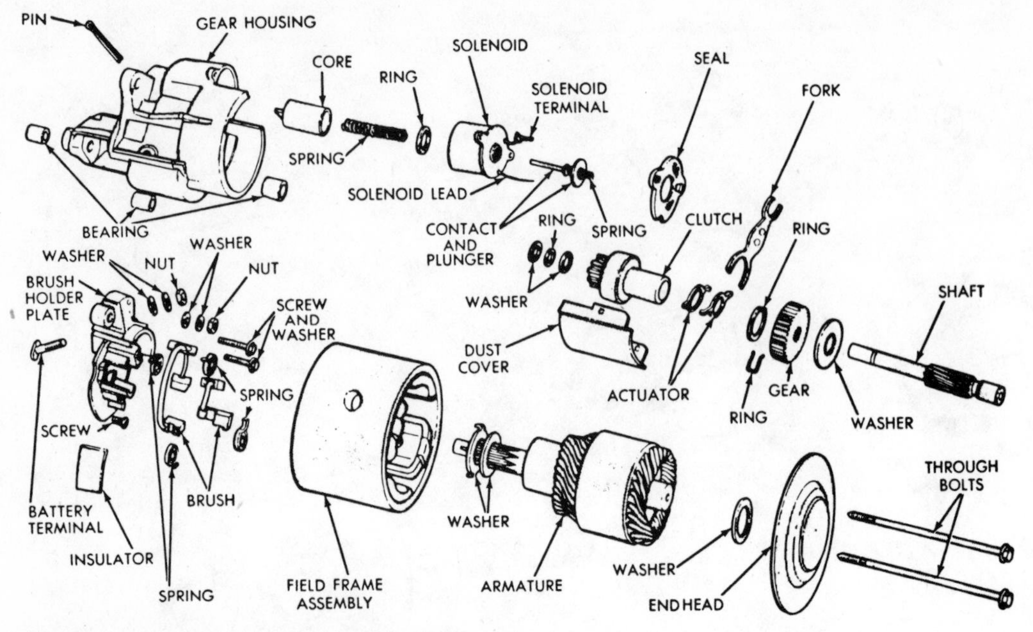

Fig. 6-9. Chrysler geared starter in exploded view.

other unfamiliar mechanism for the first time, lay the parts out on a clean bench, in order of disassembly.

Repair

Clean the starter internals with Freon, trichloroethylene, MEK, or some other approved solvent for electrical machinery. Petroleum-based solvents attack insulation and should be avoided.

The pros use a stepped driver—with the initial OD turned to shaft diameter and a flange that is slightly less than bushing OD—to extract the open-bore bushings. A hardwood dowel sized to bushing OD will work almost as well. Blind bushings, i.e., those pressed into closed bores, can be extracted hydraulically, as described with reference to pilot-shaft bearings above, or can be carefully split with a chisel and collapsed inward. Avoid nickling the bushing bore.

New bushings are pressed into place and not finish-reamed. Check shaft straightness in a drill press or by rotating the shaft on precision vee blocks. Rebuilders try to hold runout to 0.001 in. (0.002-in. total indicator runout). Some shaft wear can be tolerated, but deep scores mean that the armature should be replaced with a good used part. Most domestic starters are cheaply obtained from wrecking yards. Foreign starters are another matter, and you will save money if you purchase from a supplier of cores to the rebuilding industry.

Inspect the commutator: in most cases all that will be required is to polish the segments with 000-grade emery paper. An out-of-round armature can be sent to a rebuilder for turning and undercutting. Figure 6-10 illustrates the latter operation, as performed by hand. (Shops that specialize in this kind of work employ more sophisticated tooling.)

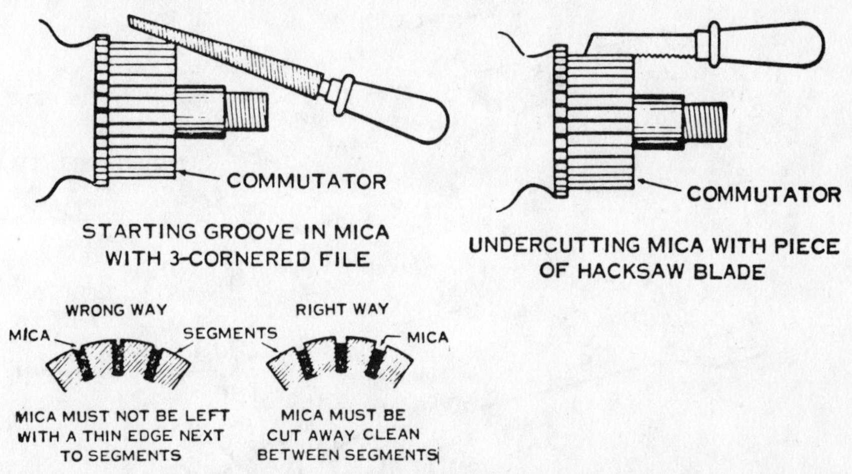

Fig. 6-10. It is necessary to undercut the "mica"—the material formerly used for commutator insulation—below the surface of the commutator bars. Otherwise, brush wear will be rapid.

Replace the brushes, making certain that brush holders are securely riveted to the frame. Two brushes are grounded; two are insulated. Soldered brush connections, e.g., certain Ford 390 units, require professional help in the form of a 500W soldering gun.

FULL-CIRCLE SEAL

Many recent U.S. and imported engines employ a full-circle, or radial, rear crankshaft oil seal, located outside of the block (Fig. 6-11). Make note of any leakage and remove the seal. Check carefully for distortion at the mounting flange, which is quite common and caused by excessive torque on the flywheel bolts. As an assembly reference, observe that the sharp edges of the seal lips are inboard, toward oil pressure.

OIL PUMP

The oil pump is a sacrificial item, replaced as part of the rebuilding process. Most owners opt for a new, as opposed to rebuilt, pump.

Even so, take a few minutes to examine the old pump. Outright failure of pump or drive rod usually is the result of debris caught between the pump case and impeller. Plastic drive rod sockets, used on Chevrolet V-8 intermediate oil-pump-drive shafts, can fatigue and *must* be renewed upon assembly. High-volume pumps are sometimes retrofitted to these engines to increase oil pressure at idle. The pressure rise is a function of the greater volume of oil delivered and the stiffer relief-valve spring, almost always incorporated into these pumps. Of course, nothing is free, and high-volume pumps put additional stress on the plastic shaft connector. Replace the OEM intermediate shaft with an all-metal aftermarket part, such as TRW PN 53010 for the small block and PN 53026 for the big block.

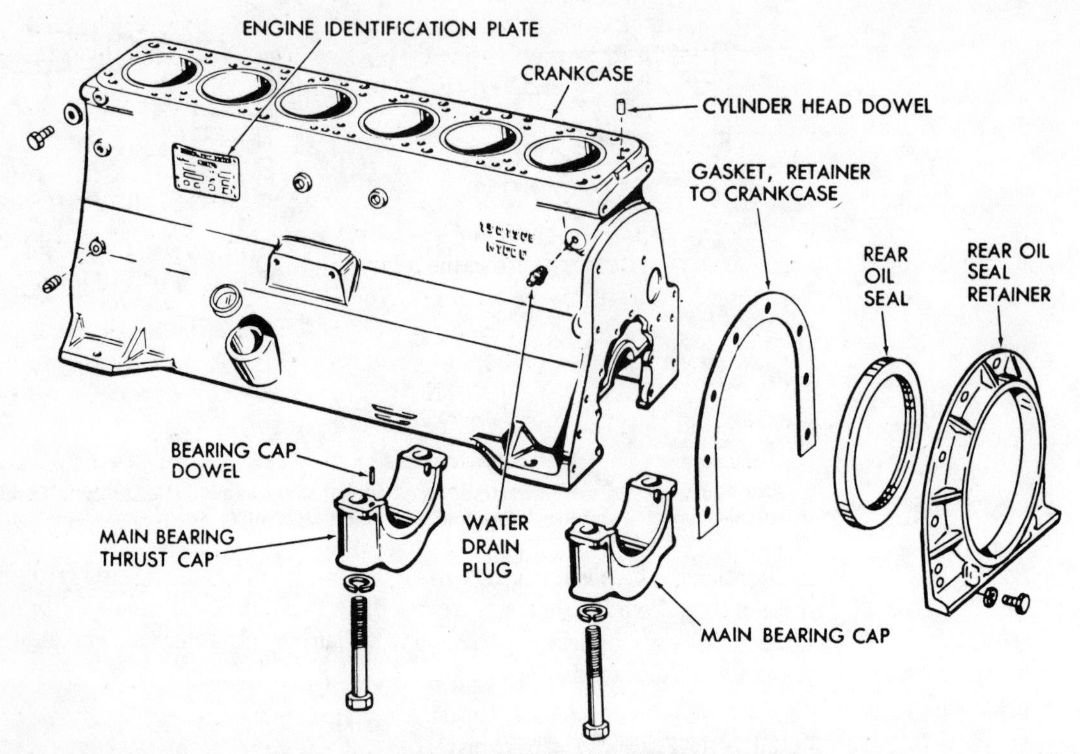

Fig. 6-11. One-piece rear oil seals were first used on industrial engines, such as this Waukesha, and European auto engines. Major problem affecting these seals is crankshaft hub distortion caused by overtightening the flywheel bolts.

Foreign material, usually in the form of nylon timing gear sloughings, can enter the pressure relief valve: if the valve is jammed open, oil pressure drops to zero; if the valve is held shut, high oil pressure can rupture the oil filter canister. Erratic or low oil pressure can often be traced to air leaks on the suction side of the pump. Inspect the pickup tube carefully, looking for loose joints and cracks.

Applications

One of the problems associated with the GM 231 engine is low oil pressure, usually encountered after major work. Part of the difficulty appears to be loss of oil pressure at the cam bearings; some reports indicate that a bearing set has a half-life of three months. Another line of attack is to use the post-1980 0.500-in. oil pump pickup tube on earlier blocks, together with some careful die-grinder work to smooth the sharp angles on the suction side of the pump. Melling Tool builds a high-volume pump for this engine, which is said to boost hot idle oil pressure from 7 psi to as much as 20 psi.

Early GM V-6 173 CID may experience sudden oil pump disengagement. The cure is to replace the drive rod with the current OEM part, which is hexed along its full length. Upon assembly, care should be taken to assure that the rod is mated securely with its retainer. Replace the OEM pump with a better-than-OEM aftermarket unit.

Chrysler 318 pumps mount to the rear main bearing cap, which should be remachined if distorted more than 0.060 in. over a 6-in. span. The pump is then installed with a spacer and 1 ½-in. ⅜-16 capscrews (Fig. 6-12).

HARMONIC BALANCER

The balancer can be a sliding fit or, as in the case of Chevrolet big block V-8s, it can be pressed on the crankshaft nose and require a hub puller for extraction. This tool bolts to tapped holes in the balancer hub. Do not use an ordinary gear puller that hooks over the balancer rim. Not only will a gear puller fail to remove a stubborn balancer; it will destroy the component in the process. Pinto ohv engines are an exception to the "no gear puller" rule: the hub on these engines is flanged for purchase (Fig. 6-13). Note the crankshaft protector, which should be used on all applications involving a puller.

Examine the balancer carefully: failure can result in main bearing damage or crankshaft breakage. The most common type of harmonic balancer uses rubber as the dampening medium. These units can be tested in with a bearing press (in which case, the machinist should do it) or with a drill press. Place the balancer on the machine table with shims under its rim, so that the hub can be displaced downward. Apply moderate pressure—no more than 100 lb.—and observe the condition of the metal-to-rubber bond. Separation or cracks mean that the balancer should be discarded and replaced with an *exact* equivalent.

Viscous balancers are more difficult to check and the final determination should be made by your machinist. However, if the unit rattles when rotated, you can be sure it has lost fluid.

Other areas to examine are the keyways on both the crankshaft and balancer hub, and the front oil seal contact surface. A loosely fitting or seal-cut balancer should probably be discarded, although the latter condition can sometimes be corrected with a sleeve and properly sized front seal (see Chapter 7).

A special note on 1985 and earlier GM 3.0- and 3.8-liter V-6s: oil leaks at the harmonic balancer attaching bolt can be cured by coating the washer with a non-hardening sealant or by substituting the current OEM piece for the original washer. Ask for PN 25525918.

FRONT COVER

Also known as the timing case or cover, the front cover usually mounts the front oil seal, water pump, and, on some engines, the oil pump and distributor. The cover may be reluctant to come free, either because it is glued in place with sealant or because of a tight fit on locating pins. But before doing anything drastic by way of pry bars, make absolutely certain that all fasteners are removed. Front covers are often secured at the bottom by pan bolts and may also be bolted to the underside of the head, Japanese fashion. Other less obvious fasteners, such as water-pump thru-bolts and the "back-of-the-oil-pump" bolt on Datsun L-series fours, may also be present. Consequently, remove all

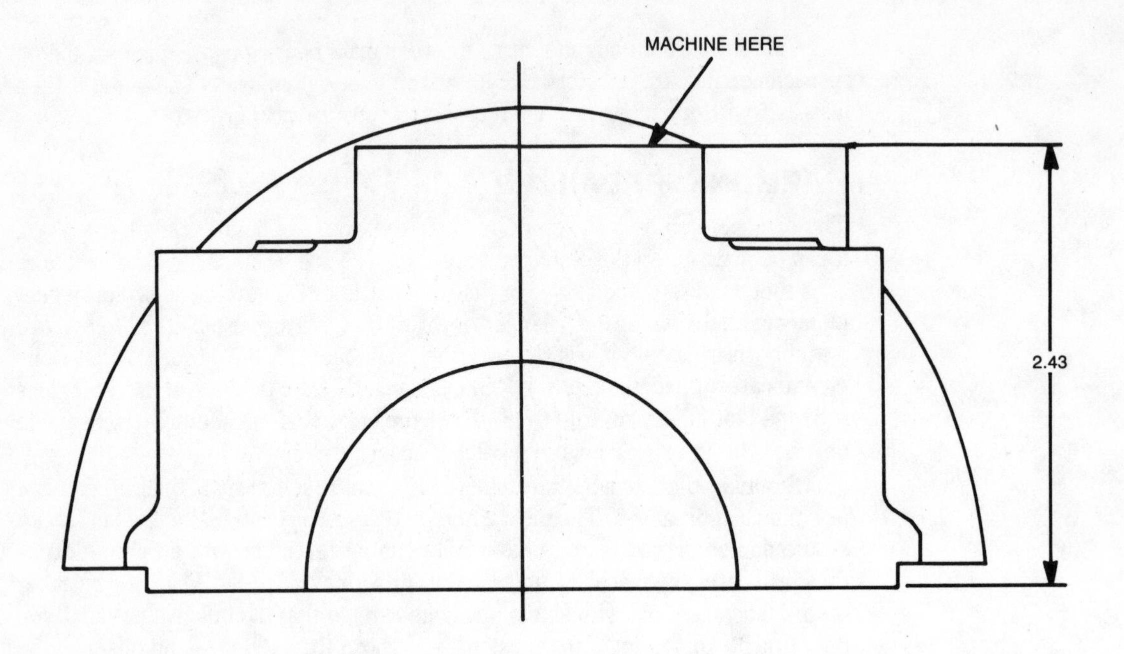

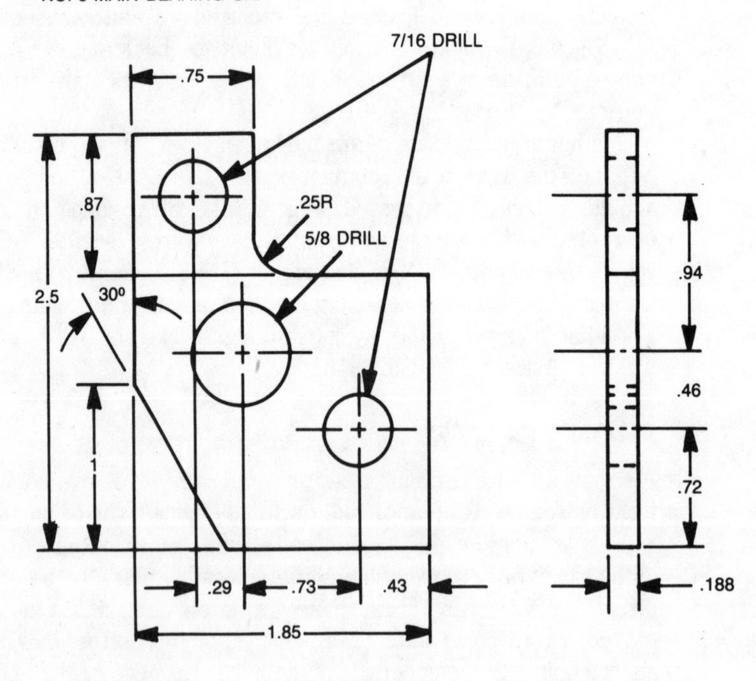

Fig. 6-12. Chrysler suggests these modifications for 318 engines that experience pump or drive rod failure because of pump-mounting surface misalignment.

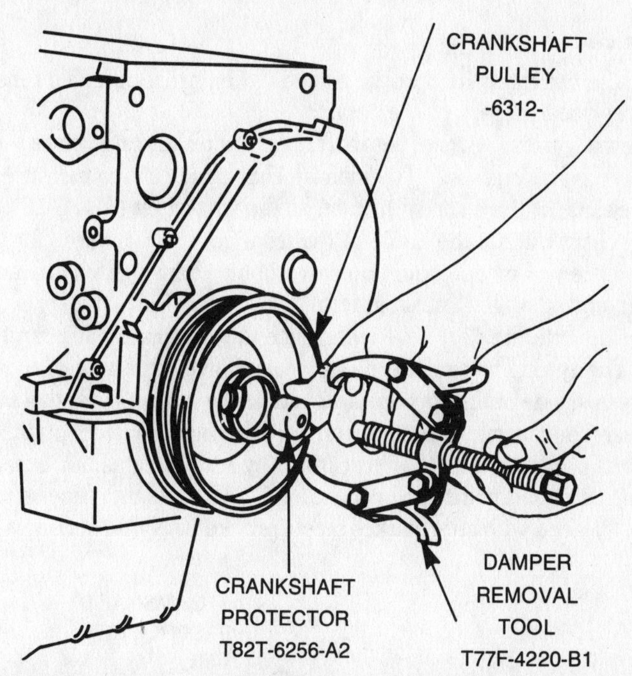

Fig. 6-13. Ford 2.3-liter balancer is removed with an ordinary gear puller. Note the crankshaft protector that can be in the form of a wrench socket or heavy washer, and which should be used with any puller.

accessory hardware from the front cover before attempting to detach the cover from the engine.

Once the cover is removed, rotate the crankshaft to tdc on No. 1 piston and make careful note of the timing marks (Fig. 6-14). This is the last time you will see these marks in their factory-set alignment.

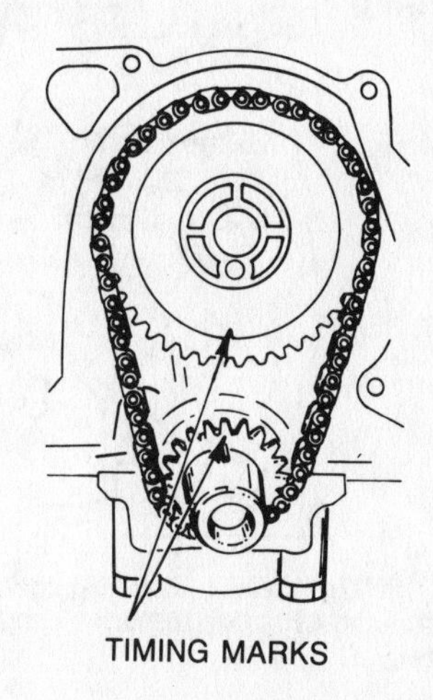

Fig. 6-14. Chrysler small-block V-8 timing marks. (NN358)

Chain Drive

Normally, the timing chain, sprockets, and (when present) chain tensioner and rails are replaced as part of the rebuild (Fig. 6-15).

There are various ways of determining chain pin and side plate wear, which translates as chain "stretch." Traditionally U.S. practice is to measure deflection on the slack side of the chain with ½ in. as the wear limit (Fig. 6-16). In severe cases, the chain will rub the inside of the front cover. Standards for ohc chains vary with make and model, but in no case should chain stretch amount to more than 2 percent of original chain length.

Sprockets tolerate very little wear since the case hardening rarely exceeds 0.020 in. The typical wear pattern is a progressive flattening of tooth profiles on the stressed side of the sprocket that ultimately gives the teeth a pronounced hook. Wear concentrated on the outboard or inboard side of the sprocket indicates chain misalignment, often caused by loss of camshaft thrust control.

Nylon-cushioned camshaft sprockets, OEM on many American engines, should be replaced with aftermarket steel sprockets. Nylon becomes unreliable with age.

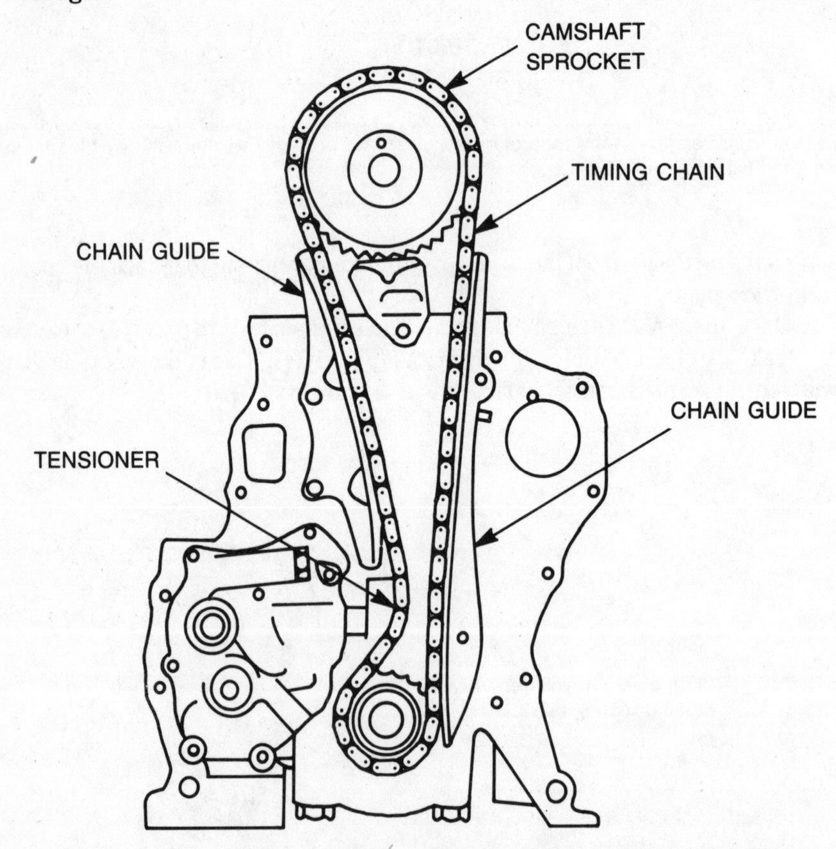

Fig. 6-15. Lightweight chain, guides, and hydraulic chain tensioner on Mitsubishi-manufactured Chrysler 156 are typically Japanese.

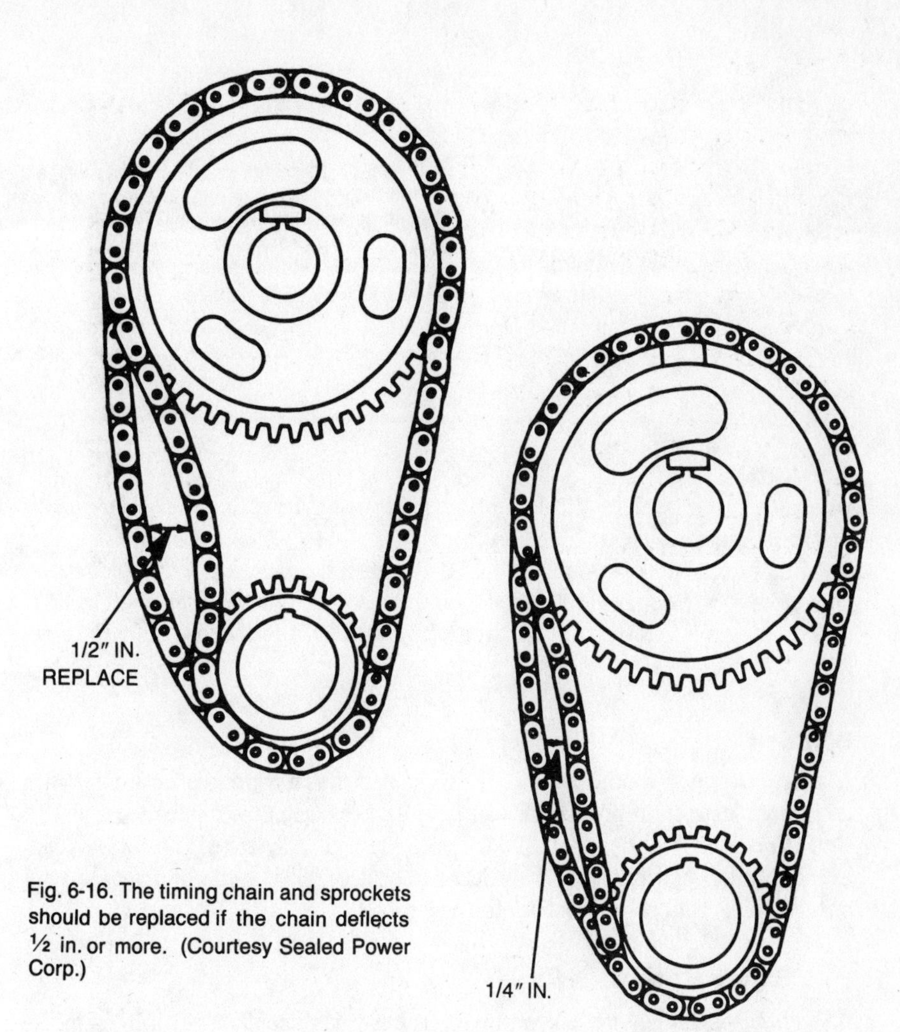

Fig. 6-16. The timing chain and sprockets should be replaced if the chain deflects ½ in. or more. (Courtesy Sealed Power Corp.)

Pressed-on camshaft sprockets are normally left to the machinist to extract; crankshaft sprockets can be pried off with a pair of screwdrivers or, should that fail, can be pulled with a bearing separator. Some shops take out sintered-iron crankshaft sprockets with a cold chisel. A sharp rap splits the sprocket in half.

Gear Drive

Noisy operation is the most common complaint, and for steel gears it is usually traceable to excessive backlash, or clearance between the gear teeth. Backlash can be measured with a dial indicator or feeler gauge (Fig. 6-17) and, as far as the Ford big block sixes (the most popular engine using this type of cam drive) is concerned, the wear limit is 0.004 in.

Catastrophic failure usually takes the form of severe wear, which can include surface pitting, loss of teeth on account of fatigue failure, and case crushing,

Fig. 6-17. Backlash can be checked with a dial indicator or feeler gauge. Excessive backlash means noisy operation; insufficient backlash can overpressure the gears and lead to camshaft bearing failure. (Courtesy Sealed Power Corp.)

where the case hardening flakes off in slivers. Damage to either gear means that both must be replaced.

Fiber camshaft gears, most often encountered on vintage Chevrolet sixes, should be renewed, regardless of apparent condition. The crankshaft gear in these engines is subject to little wear, and usually it is merely enough to replace the fiber gear.

Belt Drive

Gilmer, or cogged-tooth, belts should be replaced as a routine precaution, since no known inspection procedure can give early warning of failure (Fig. 6-18). Belt life can be as little as 50,000 miles. Failure usually takes the form of loss of tooth at the belt/crankshaft sprocket interface. This condition may take out one or more pistons and associated valves.

PISTON ASSEMBLIES

As mentioned earlier, an engine up for rebuild has usually worn ridges in the cylinder bore at the upper limit of ring travel. Deep ridges "snag" the rings and must be reamed before the pistons can be extracted. Pistons that will not be reused can sometimes be forced past the ridges, saving the cost of renting or buying a ridge reamer.

Dismantling

Figure 6-19 illustrates a professional ridge reamer. A cam and nut arrangement forces the carbide blade against the bore and the tool—and like all other reamers—is turned counterclockwise to cut. Lubrication is optional. When adjusted and sharpened properly (a problem with rental units), the reamer makes a smooth cut, flaking the metal without leaving scores. Insufficient pressure allows the blade to skate—polishing, rather than reaming. Excessive pressure causes the tool to judder and can splinter the carbide blade.

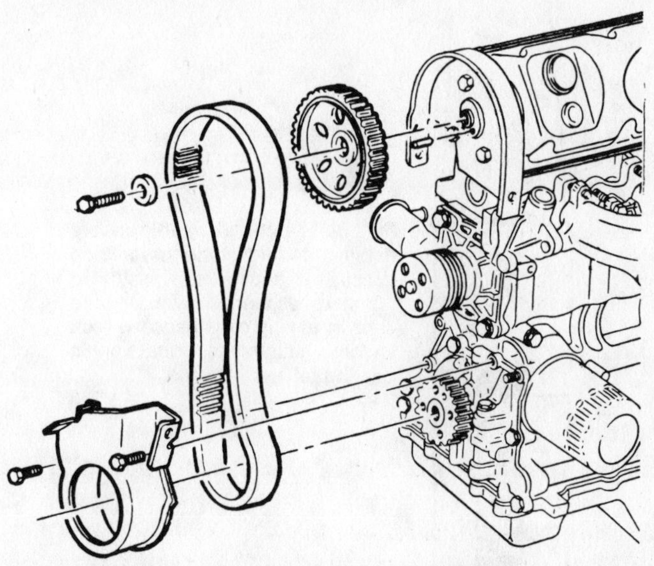

Fig. 6-18. Gilmer timing belts have advantages over chains: belts are silent, need no lubrication (in fact, would be destroyed by it), and are inexpensive at the OEM level. But belt reliability was a problem for the Vega pictured here, and continues to be less than ideal for more modern engines that use this form of camshaft drive. (Courtesy Chevrolet Div., General Motors Corp.)

Do not attempt to remove every last vestige of the ridge. That is practically impossible and damaging, since extremely deep cuts will be required to compensate for bore distortion. All that is necessary is to get the pistons out; final bore preparation, including removing what is left of the ridge, will be done by the machinist.

Rotate the crankshaft to bring No. 1 piston to bottom dead center. Examine the flanks of the connecting rod and cap for marks. Both the rod and cap should be stamped "1" (Fig. 6-20). If this mark is absent (e.g., small block Chevrolet), difficult to decipher (Ford), scrambled with any number between 1 and 8 (Oldsmobile), or if rod and cap numbers do no match (the result of a

Fig. 6-19. Some auto parts stores rent ridge reamers, hub-type gear pullers, piston ring compressors, and other specialized tools. The ridge reamer shown is a professional-quality tool, supplied to Kabota Tractor mechanics.

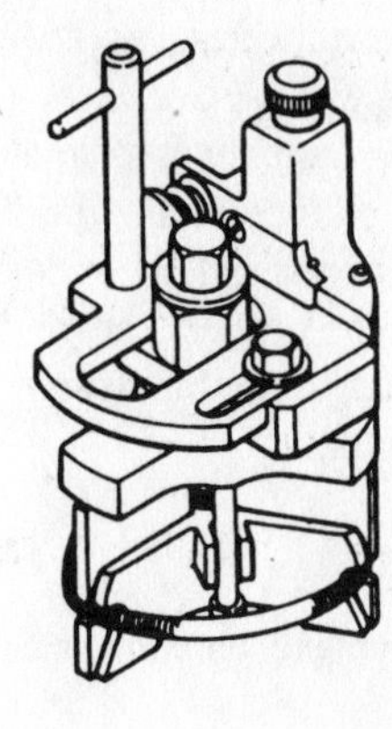

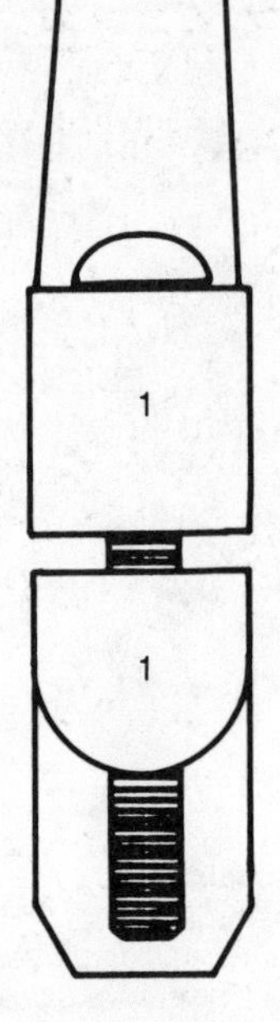

Fig. 6-20. Both rod and cap are marked and must be assembled exactly as found with matching numbers aligned. Reversing the caps or mixing caps between rods almost guarantees catastrophic bearing failure.

clumsy overhaul), make your own identification marks. Using a pin prick, stamp one dot on the cap and one dot on the adjacent rod surface—two dots for No. 2, three for No. 3, and so on. Make a written note of the position of these marks relative to some prominent engine feature such as the camshaft or oil pump. Otherwise, the piston assemblies could be installed 180 degrees out of phase. While the engine will run with one or more pistons reversed, severe knock will result.

Remove the rod nuts, backing them off in even increments. Rock the crankshaft a few degrees to separate the rod and cap; if the cap is stubborn, separate it by driving the piston deeper into the bore. A light tap on the underside of the piston with a hammer handle should be enough to release the cap. Slip the cap off its bolts and set it aside, with bearing shell still in place.

You may wish to install crankshaft protectors on the rod bolts before continuing. Commercial, screw-on protectors can be purchased, but a short length of fuel line pushed over the bolts' threads works as well.

Using the hammer handle as before, drive the piston out of the bore. Do no allow the pistons to hit the floor.

With the upper and lower bearing inserts in place, reinstall the cap and nuts. Marks should be properly aligned and the nuts run down with their bright sides against the cap. Hang the assembly up on a nail and repeat the process for the remaining pistons.

INSPECTION

A common type of damage is erosion of the piston crown edges, caused by detonation or pre-ignition (Fig. 6-21A). The piston will appear nibbled, as if a

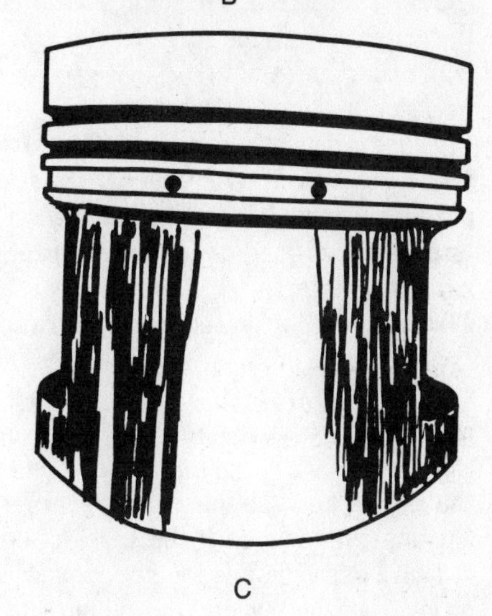

Fig. 6-21. Detonation and preignition nibble away at the edges of the piston (A); lack of lubrication or lugging scuff the major thrust face, which is defined by the slight offset of the piston pin relative to the piston centerline (B); and scuffing near the piston pin bores points to a tight piston pin (C).

metal-eating mouse were at work. In severe cases, the piston crown will be holed near the centerline, blasted by the impact of detonation, or burned and melted by preignition. The latter can result from any condition that produces local overheating (cooling system deposits, usually associated with leading or tailing cylinders, hangnail spark plug threads, partially detached flakes of carbon that heat to incandescence).

Figure 6-21B shows a dry scuff, confined in this case to the thrust side of the piston. The problem is caused by lack of lubrication and/or lugging. If scuffing occurs on both sides of the piston pin (major and minor thrust faces), the diagnosis is similar, but detonation can be added to the list of possible causes.

Scuffing at a 45-degree angle to the piston pin means that the swing of the rod has been curtailed by a tight piston pin (Fig. 6-21C). Look for failed bearing surfaces on both the little end of the rod and the piston bosses.

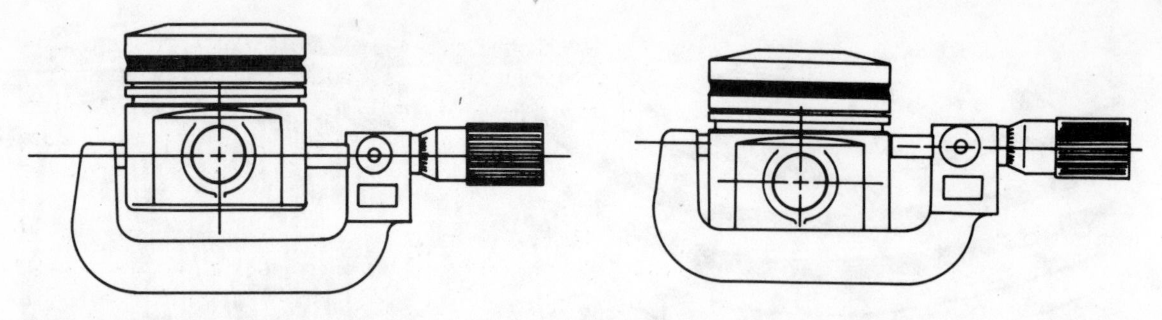

Fig. 6-22. Piston diameter varies with the point of measurement. Most Sealed Power automotive pistons are measured at the top of the thrust area, just below the lower ring groove. Heavy-duty pistons from the same manufacturer are measured 180 degrees out from the piston pin centerline.

Pin lock failure, signaled by erosion of metal near the pin holes, is often the result of a bent connecting rod or crankpin taper. Pin lock failure can send the piston pin into the bore and effectively destroy the block. Snapped pin locks, with no evidence of associated groove damage, indicates that the locks were overstressed during installation.

Piston diameter is measured across the thrust faces (at 90 degrees to the piston pin). If you wish to verify the measurement or to check the fit of a new piston, you must follow the manufacturer's instructions. Pistons are cam-ground, or egg-shaped, and tapered toward the crown. Some manufacturers measure from the centerline of the piston pin, others at a point ¾ to 1 in. above the skirt (Fig. 6-22). All measurements are made at room temperature (70 degrees F.).

Knurling will restore thrust faces to specification, but the fix is temporary and hardly worth the effort.

Wear may also appear as overly wide compression-ring grooves, which allow the ring to twist as the piston reverses direction. This condition is known as "pound-out," although the problem begins with frictional wear. Rings move radially, expanding and contracting within their grooves, and rotate. As the groove becomes wider, pound-out develops and rapidly erodes the groove, sometimes breaking the land between the first and second compression rings. Detonation has the same effect.

Worn piston grooves can be cut oversize and brought back to spec with steel inserts. However, the repair is hardly worth the effort for light vehicles.

Select replacement pistons from a reputable and preferably OE manufacturer, such as Federal Mogul, Imperial Clevite, Perfect Circle, Sealed Power, TRW, or Zollner. Off-brand pistons can be a world of trouble, particularly if the oversized replacements are heavier than the originals. This plays havoc with engine balance.

Most pistons are cast from eutectic or aluminum alloys such as SAE 332, SAE 334, or F-132. "*Eutectic*" and "*hypoeutectic*" means that silicon, zinc, and other minerals are finely divided throughout the base aluminum—like vanilla ice cream. Such alloys are easy to cast and machine, and they perform satisfactorily in most applications.

Turbo-charged, Z-28, and other high-output engines generally require forged pistons. Forging eliminates voids in the metal and compacts the internal grain structure at the crown, pin bosses, and ring lands. These pistons withstand heat better than cast pistons, but are expensive to replace and noisy on cold start-up. Typically, a forged piston needs 0.002 in. more bore clearance than a cast piston. TRW is the major aftermarket source.

Zollner recently developed a proprietary hypoeutectic alloy for cast pistons. Clusters of long-wearing silicon are spotted throughout the metal. Zollner's has the same relationship to eutectic alloys that Irish stew has to thin gruel. The new alloy is claimed to have superior fatigue strength, less thermal expansion (so that pistons made of it can be set up tighter), better resistance to scuffing, and superior wear properties. Forged pistons are stronger than Z-16 when cold; but tensile strength falls off rapidly with temperature. At 330 degrees F., Z-16 is clearly superior to conventional forged alloys. For OEM applications, Z-16 pistons can reduce piston mass by as much as 25 percent.

For rebuilders, the good news is that Z-16 pistons are available for some Ford applications, including Ford 1.9 EFI and 3.0 liter engines, and for 302 high-output engines built from 1984 to present.

Rings

Abrasive wear leaves a dull, satin surface on the rings, almost as if they had been anodized (unfortunately, this surface neither photographs nor reproduces well). Chrome flashing on compression rings will be eroded through to the base metal and expander humps on the oil rings may contact the cylinder walls. Run your fingertips lightly over the cylinder bores: the surface should be glazed and ice-smooth. If the bore feels like a cat's tongue, you can be sure that the air cleaner has long since failed. Cylinders should be bored to the next oversize and air leaks around or below the air cleaner must be found and corrected.

Scuffing occurs when local hot spots transfer metal from the ring surface to the cylinder walls. The scuffed area appears rough to the eye and, when viewed under a microscope, is reminiscent of a lunar landscape. Basically, the condition comes about as a result of overheating, but the exact cause can be elusive. Engineers at Sealed Power suggest these possibilities:

Symptom	Possible Cause
Overheating	• Clogged or restricted cooling system (if damage is limited to one or two cylinders, look for local waterjacket restrictions) • Defective thermostat • Loss of coolant • Water pump failure • Detonation

Symptom	Possible Cause
Lubrication failure	• Worn main bearings • Oil pump failure • Engine lugging under load

	• Prolonged idle • Low oil level • Coolant in oil • Failure to pressurize system after rebuild
Symptom	**Possible Cause**
Wrong cylinder finish	• Low crosshatch finish, allowing rings to run dry • Failure to hone cylinders after reboring
Symptom	**Possible Cause**
Insufficient clearance	• Inadequate bearing clearance at either end of rod • Improper ring size • Cylinder bore distortion

Ring breakage is due to abnormal loads or localized stresses. The most common cause is ring sticking from varnish and carbon deposits. Part of the ring binds to the groove, while part flutters like a banner. Detonation accelerates ring groove pound-out and, in severe cases, shatters the compression rings. Other possible causes include excessive cylinder taper, which usually breaks rings near the gap area, and overstressing during installation, which breaks the rings 180-degrees out from the gap. Oversized rings, sometimes mislabeled by the manufacturer, butt their ends and shatter into slivers.

Ring wear, or loss of thickness, is academic for an engine that will be rebuilt. However, it is worth knowing that wear can be estimated by measuring the end gap, as described in the next chapter. Make the measurement with the ring positioned near the bottom, and least worn, part of the cylinder bore. No. 1 compression ring wears fastest, followed by No. 2. The oil ring may show little or no wear. For a number of years, Harley-Davidson chrome-plated their oil rings. The practice was stopped when it was discovered that the plated rings outlasted every frictional surface in the engine.

Materials

Most engines use plain iron compression rings, flashed over with a light (0.004 in.) coating of chrome. Besides providing a hard, abrasive-resistant wearing surface, chrome develops microscopic cracks which act as oil reservoirs. Typically about 2 percent of ring surface will be voided.

On the debit side, chrome rings are notoriously slow to seat. Some oil burning can be expected for the first few hundred miles of operation. A more serious disadvantage is that chrome rings are incompatible with "soft" blocks. When in doubt, use the OEM-type ring; otherwise the block may wear faster than the rings.

Molybdenum rings were introduced in the late 1950s and were immediately recognized for their ability to reduce scuffing. The outer diameter of the ring is grooved, and moly is sprayed on with a plasma or other bonding process. Besides having a very low coefficient of friction and a very high melting temperature (4750 degrees F.), moly gives a piston ring surface that is 15 to 30 percent void. Initially, researchers credited the metal's high melting point and the ample reservoir space for moly's resistance to scuffing. More recently it was discovered that moly forms oxides which act as solid lubricants at temperatures above 1050 degrees F.

This discovery led TRW to develop what the company calls the Plastic Ceramic Piston Ring. These rings are face-coated with a mixture of aluminum and titanium oxide applied at a temperature of 30,000 to 40,000 degrees F. After machining, the porous coating is 0.004 to 0.008 in. thick. According to TRW, this coating acts as a dry lubricant and resists scuffing better than moly. Detonation, which causes chunks of moly to fall out, has little effect on the ceramic inlay. Abrasive resistance is said to be superior to that afforded by chrome rings.

Recommended side clearance is 0.002 to 0.004 in. If these rings must be narrowed, grind on the top, or the nonsealing side. End gap follows OEM specifications; TRW supplies ceramic rings in both Standard Gap and File Fit configurations, which are 0.005-in. oversize to allow the installer to custom fit the end gap. Bore the cylinders to within 0.002 or 0.003 in. of desired diameter and, using a 400-grit stone, hone to fit. Cylinders worn no more than 0.001 in. can be fitted with ceramic rings after glaze breaking (see the section by that title below).

CONNECTING RODS

The machinist will disassemble each rod and piston, and should check the bearing bores on both parts. Big-end bores elongate as shown in Fig. 6-23A, a condition that interferes with heat transfer and may allow the insert to spin, with disastrous results. Your machinist can correct rod bearing stretch by trimming the cap/shank parting face (Fig. 6-23B) and then honing, grinding, or boring the big end to specification (Fig. 6-23D). Boring is fast; grinding gives a uniform finish on different materials; but honing reduces the amount of cap metal that must be removed. From the owner's point of view, it is the most desirable way to recondition the rod. Lost cap metal translates as loss of compression ratio.

Magnifluxing is optional, but highly recommended on vintage engines, high-performance engines of whatever date of manufacture, and any engine that will see severe service (Fig. 6-24). To the author's knowledge, no one has compiled statistics on conn rod failures in lightweight engines, but it is apparent that rod cross-section and bearing area have been reduced, and prudence would suggest that these late V-8 and V-6 rods be checked for fatigue cracks.

Complete rod refurbishing means checking the small-end bores, installing new small-end bushings (when applicable), checking rod straightness, and polishing rod flanks to remove small nicks. Bent rods, checked between big-end and small-end bores, are traditionally cold-straightened. This process is by no means ideal, but has worked since the earliest days of the industry. The rod, weakened by the initial bend and subsequent straightening, is prone to bend again. Any rod so treated should be Magnifluxed prior to assembly.

In general, rod nuts should be replaced. Rod bolts rarely fail on street engines, but if bolts are to be replaced, use the crankpin as a work support.

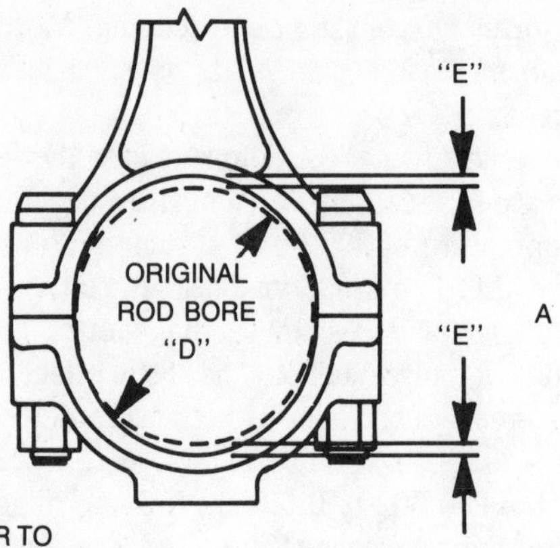

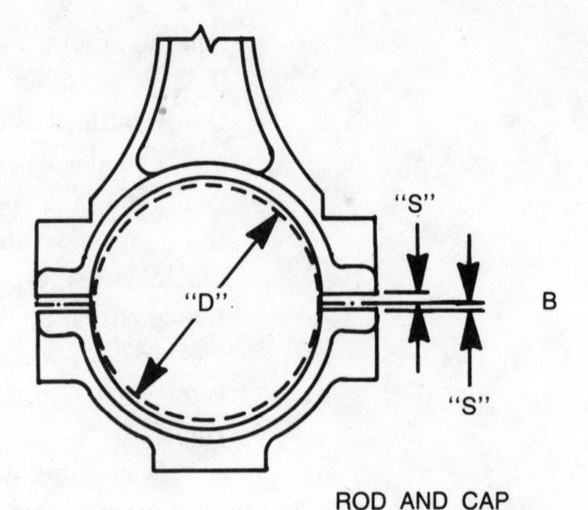

ROD PRIOR TO RECONDITIONING "D" SHOWS TRUE-ROUND BORE OF ORIGINAL ROD. "E" IS AMOUNT OF STRETCH AT TOP AND BOTTOM.

ROD AND CAP SHOWN IN SAME RELATIVE POSITION AS ABOVE, AFTER GRINDING OF PARTING FACES. "S" INDICATES AMOUNT OF STOCK REMOVED FROM ROD AND CAP.

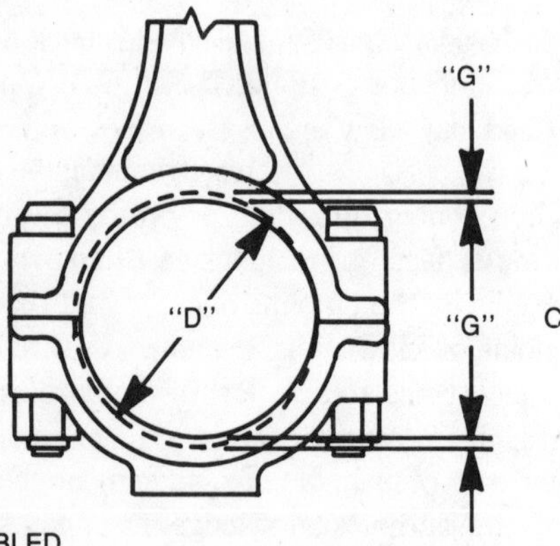

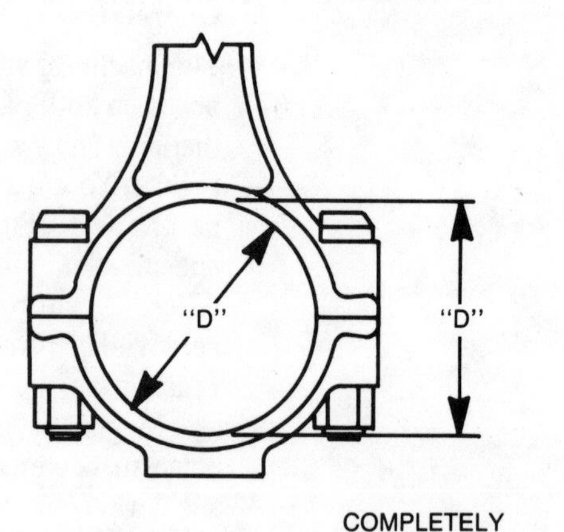

RE-ASSEMBLED ROD NOW HAS A SMALLER VERTICAL DIMENSION (BY AMOUNTS "G") THAN DIAMETER "D" OF ORIGINAL TRUE-ROUND FACTORY ROD.

COMPLETELY RECONDITIONED ROD AFTER HONING. ROD BORE IS AGAIN TRUE-ROUND, AND OF THE SAME DIAMETER "D" AS THE ORIGINAL FACTORY ROD.

Fig. 6-23. Elongated big-end bores can be corrected by trimming the rod and cap parting faces and machining the bores to size. (Courtesy Sunnen Products Co.)

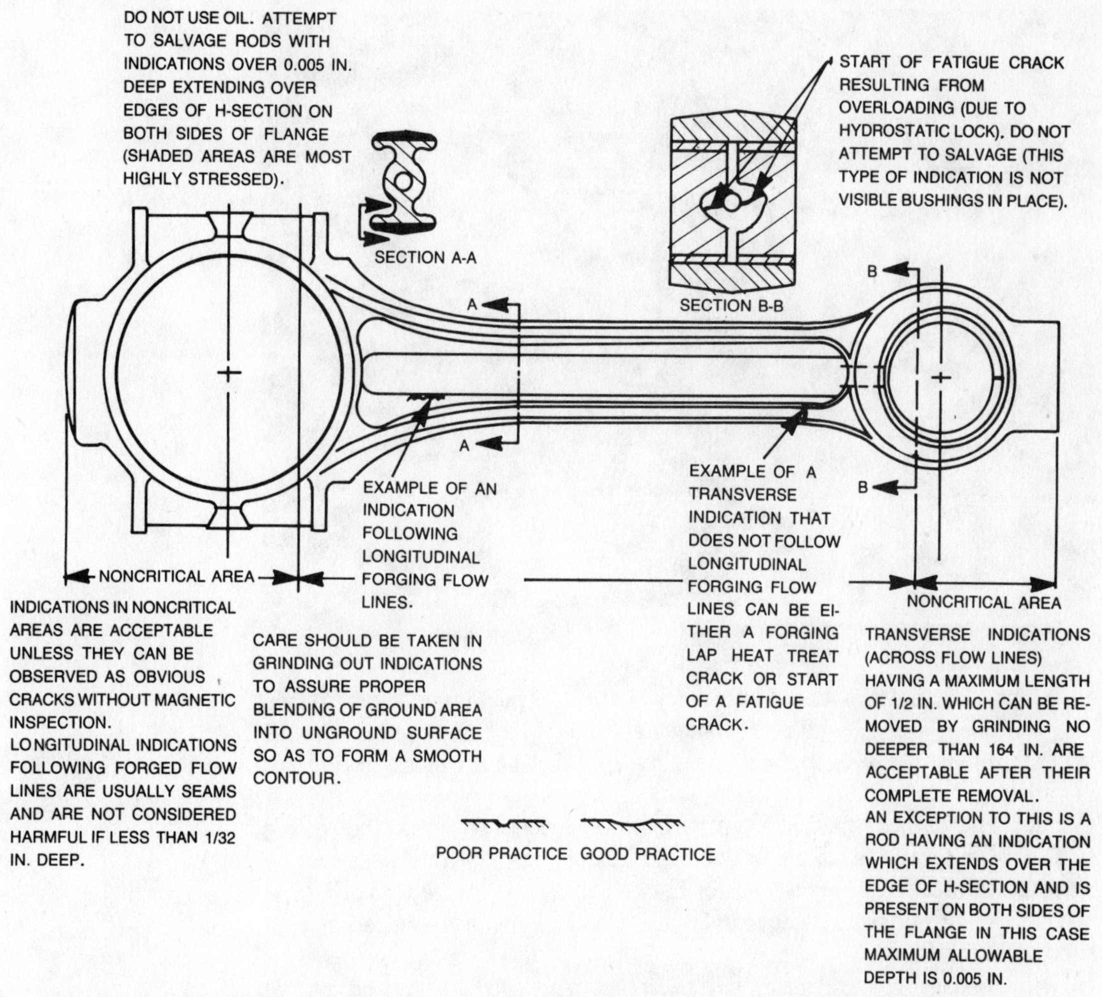

Fig. 6-24. Magnifluxing the connecting rods is good insurance for any engine. (Courtesy Detroit Diesel Allison)

That is, install the rod on the crankshaft, torque one rod nut to half-spec, and, using a punch, drive out the remaining bolt.

CRANKSHAFT

Take a careful look at the main bearing caps. All caps should be embossed with a forward-facing arrow or some other directional indicator, and all caps, with the possible exception of the first and last, should be numbered (Fig. 6-25). Those that do not carry a number can be identified by their distinctive shapes. Make your own marks if the cap marks are illegible.

Working from the center main out in both directions, progressively loosen the cap bolts, making two or three circuits before torque relaxes. Remove the bolts.

Fig. 6-25. Main bearing caps are identified by number and orientation.

Caps are usually mortised into the webs with a mild interference fit and may be further located with pins. Using a soft metal tool, pry the cap upward to disengage it. Excessively tight caps, the kind that batter metal to release, can indicate overheating.

Carefully lift the crankshaft and place it on wood blocks to protect the bearing surfaces. Deep scores, discoloration, and bearing-surface roughness can be detected by eye. A discolored—blue or black—journal is evidence of overheating and almost guarantees subsequent fatigue failure. Odds are stacked against successful repair.

Precision gauges are necessary to detect taper, flat spots, and loss of diameter. The machinist will make these measurements and grind the crankshaft as required. Many cranks will tolerate 0.030 in. before surface hardness becomes problematic, but this rule is not universal. For example, Ford V-6 232 cranks have a deep rolled main journal radius that will be compromised if journal diameter is reduced more than 0.020 in. Figure 6-26 illustrates the properly rounded ra-

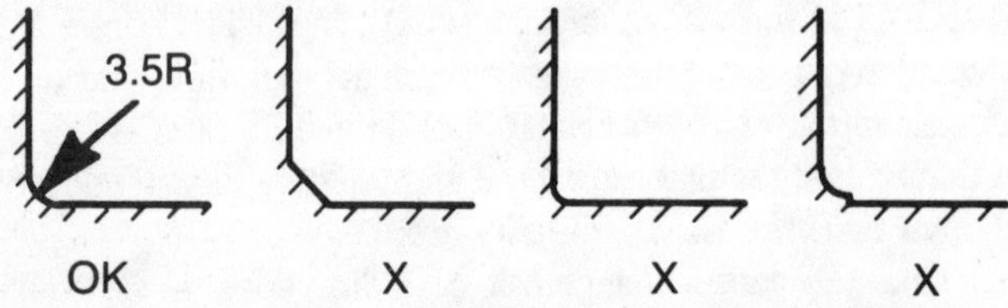

Fig. 6-26. Fillet profiles. The exact radius depends upon OEM specification, but abrupt changes in direction act as stress risers and may lead to early shaft failure. Insufficient radius clearance destroys bearings, as illustrated in Fig. 6-32. (Courtesy Marine Engine Division, Chrysler Corp.)

dius that, according to some OE manufacturers, should be achieved by rolling, rather than grinding. However, grinding is the norm and, when done correctly, does not seem to affect the fatigue life of the shaft.

Undersized journals can be built up by welding or by a process variously known as thermal spray, metal spray, or spray welding. Most well-equipped shops opt for the latter approach, since temperatures are lower and there is no danger of changing the physical characteristics of the base metal. The operator first roughens the damaged part with a grit blaster to give ''tooth'' to the coating. He then sprays the part with successive layers of powdered metal, heated to the plastic stage with an oxyacetylene flame. The part is then machined to size. At least one shop in Houston uses this technique on all crankshafts, ranging from heavy-duty diesels to single-cylinder utility engines. However, it will not work on two-cycle engines that run needle-bearing rods directly on the crankpin.

Bearings are available in standard oversizes of 0.010, 0.020, and 0.030 in. The situation becomes a bit more complex when the crankshaft is worn slightly, but not enough to justify regrinding. In this case 0.001- or 0.002-in. oversizes are used on main and/or crankpin journals. There is an additional possibility with main journals: two different-sized bearing shells can be run on the same main journal to obtain the desired running clearance. For example, all main bearings can be 0.001-in. oversize, or one or more main bearings can be made up of a 0.002-in. upper shell in combination with a 0.001-in. lower shell, or a 0.001-in. upper bearing shell and a 0.000-in. lower shell. The smaller ID bearing shell is always at the top and the variation between bearings is never more than 0.001 in. To reiterate, this procedure, sometimes called selective fitting, applies to main bearings only. It is not permissible to combine different-bearing IDs on crankpins.

Some foreign manufacturers, notably Honda and Toyota with the 1AC, 3AC, 22R, and 5M-GE engines, carry the principle of selective fitting further. Bearing ID is matched to journal OD and bearing OD is matched to connecting-rod and main-bearing bore ID. An alphanumeric code, used to guide bearing selection, is stamped on the crankshaft webs. In addition, bearing shells are color coded (unfortunately the dye washes off in service, but the coding does simplify ordering replacement bearings). OEM bearings must be used on these engines, unless conn rod and main bearing saddle bores are resized.

The machinist will check the crankshaft for straightness, supporting it in the main bearing saddles or in a special jig (Fig. 6-27). Between 0.001- and 0.002-in. total indicator runout is acceptable. Cranks that have not been Tufftrided can be cold-straightened, although the preferable procedure is to grind the crank true. The machinist should also remove the expansion plugs and mechanically clean the oil passages. Finally the crankshaft is polished to extend journal life.

Production shops rarely bother to magnetically inspect a passenger car or light truck crankshaft for fatigue cracks. The better shops Magniflux twice—before the crank is ground and afterwards. The process is not expensive and could save the engine. Figure 6-28A illustrates what to look for.

Tufftriding, a process of hardening the journals, extends crankshaft life but, unless the crank was originally so treated, is considered a luxury.

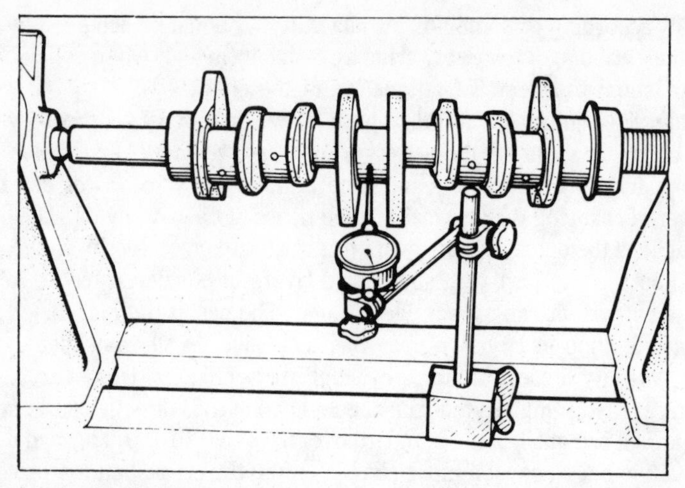

Fig. 6-27. Crankshaft straightness is checked with a dial indicator. (Courtesy Chrysler Corp.)

As indicated above, crankshaft grinding is a demanding craft with few real masters. If you have some reason to suspect the quality of work—if, for example, bearing clearances do not fall uniformly in the middle of the specification range, or fillet radii seem sloppy, or if expansion plugs appear not to have been disturbed, it may be worthwhile to test crankshaft hardness. Rapid metal removal can overheat the crank, destroying its temper. The damage is not obvious unless the crankshaft is etched.

Begin by scrubbing the bearing journals with scouring powder and water. Rinse with water, followed by alcohol. Apply Tarasov Etch No. 1—a preparation made up of four parts nitric acid and 96 parts water—to the journals with a cotton swab. Let stand for 15 seconds. Rinse with water and dry. Apply Etch No. 2--two parts hydrochloric acid, 98 parts acetone—and let stand for 15 seconds. Rinse with alcohol and blow dry.

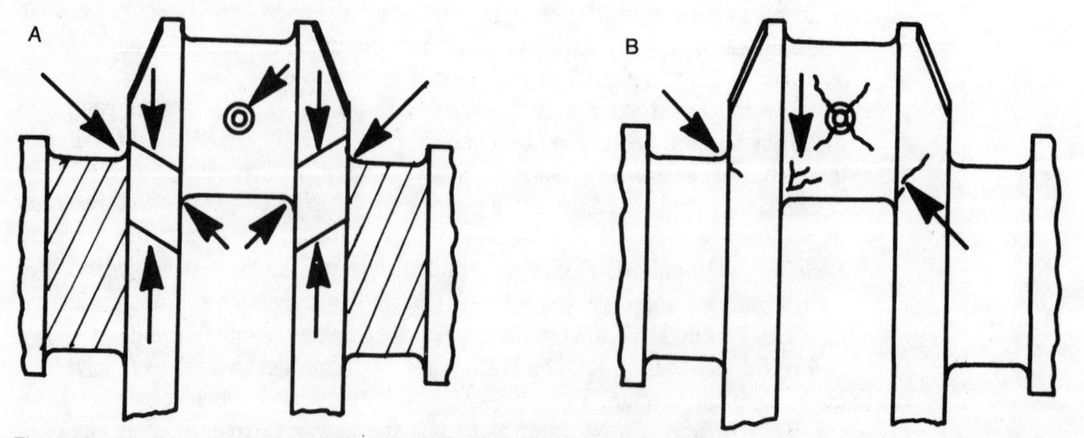

Fig. 6-28. Heaviest loads pass through the shaded areas in directions shown by the arrows (A). Fatigue cracks in these areas are fatal (B). (Courtesy Detroit Diesel Allison)

Undamaged journals will be uniformly dark gray. Softened areas will be darker, shading into black. Rehardened areas will be harder than the rest of shaft and may appear white. Any discoloration is grounds for rejecting the part. If your machinist is in the mood to take advice, suggest that he use a softer wheel, reduce feed rate, or increase grinder spindle speed.

BEARING FAILURE ANALYSIS

An examination of the crankpin, main, and thrust bearings can be quite informative and may prevent subsequent failure. Main bearings outlast crankpin bearing by about three to one, but No. 1 main, which is loaded by belt-driven accessories, may knock with the bearing clearance on the wide side of the normal range.

Thrust bearings, in the form of half-moon washers or flanges integral with the insert, control crankshaft and play (Fig. 6-29 A and B). Standard clearance is between 0.004 and 0.007 in. Thrust bearing wear is usually more severe on vehicles with standard transmissions (disengaging the clutch forces the crankshaft forward), but can also be a problem on some vehicles with automatic transmissions. Unfortunately, the diagnosis is not certain. Investigators have implicated the torque converter, shaft splines, and transmission oil pump. Any information readers have on the subject will be appreciated.

The bearing consists of as many as five distinct layers of lining material over the steel back. Figure 6-30 illustrates a Federal-Mogul overplated copper-lead bearing, considered to be the state of the art for automotive service. This bearing is identified by the letters CP, CPA, or CPB after the part number. Michigan DA-49 and Michigan-77 (formerly Clevite-77) employ the same 75 percent copper, 24 percent lead and 1 percent tin overplate.

Typical modes of crankpin bearing failure, which also apply to main journal bearings, are shown in Fig. 6-31.

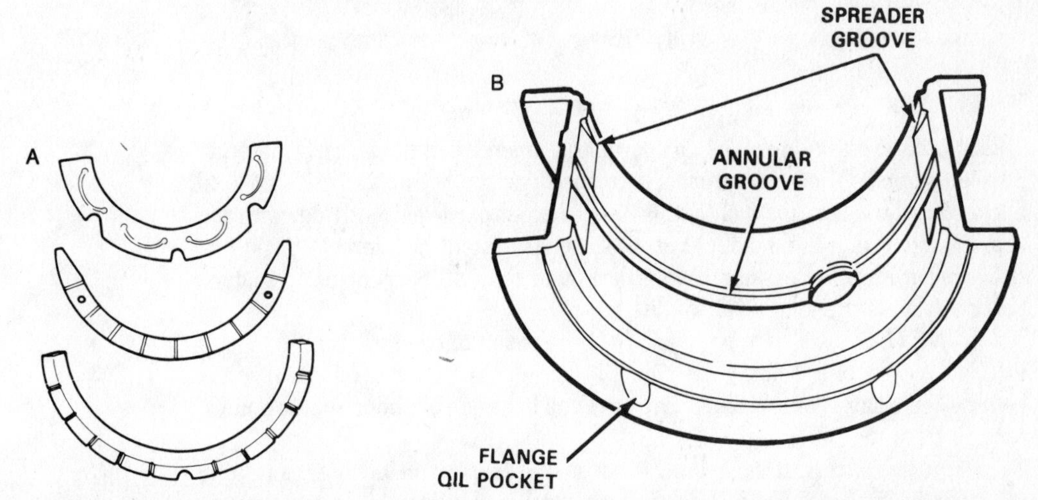

Fig. 6-29. Typical thrust bearing washers (A) and flanged thrust bearing (B). Annular and spreader grooves distribute oil over the journal. (Courtesy Federal/Mogul Corp.)

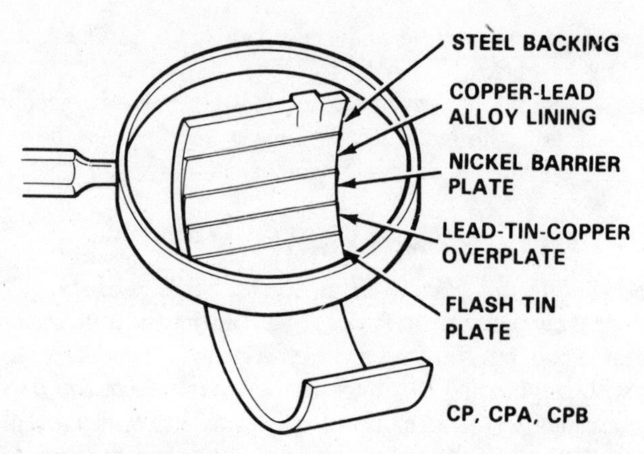

Fig. 6-30. The finest high performance automotive bearing currently available combines a lead-tin-copper overplate with a copper-lead alloy lining. (Courtesy Federal/Mogul Corp.)

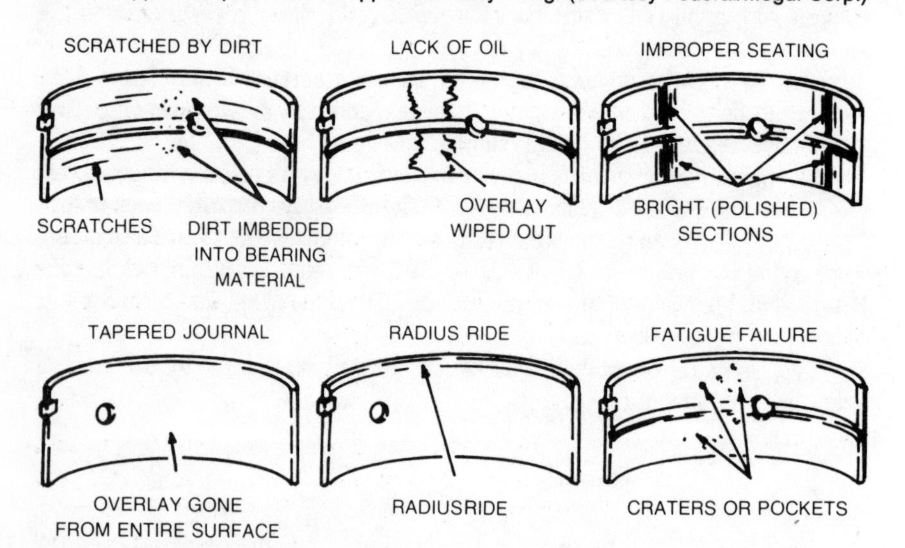

Fig. 6-31. Typical bearing failure modes. (Courtesy Ford Motor Co.)

Scratches: often caused by dirt particles embedded in the bearing material. Smaller particles tend to disappear into the overlay, but at the price of local thickening of the bearing wall (Fig. 6-32A). Larger particles protrude, with the bearing acting as a tool holder (Fig. 6-32B). As a point of interest, engineers at Michigan Engine Bearings report that more than 43 percent of premature bearing failures are the result of dirt.

Wiped Overlay: This type of failure is the result of oil starvation. If local—that is, if only one bearing is affected—suspect a plugged oil passage. A generalized oiling system failure may affect all bearings or those that are more remote from the pump.

Improper Seating: If the bearing is not seated properly in its bore, all bearing wear will be concentrated at brightly polished "high spots." The usual cause is dirt trapped between the back of the bearing and the bearing bore.

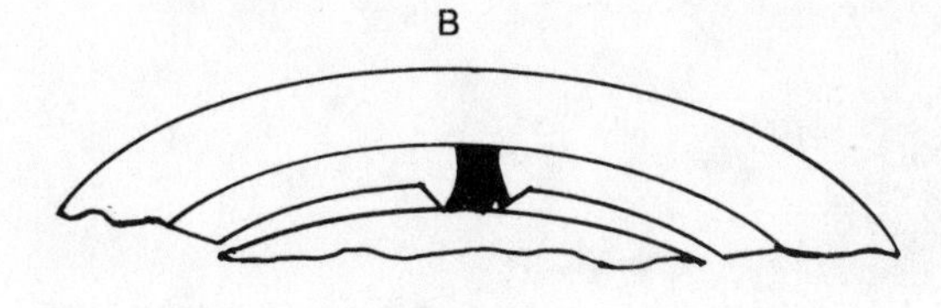

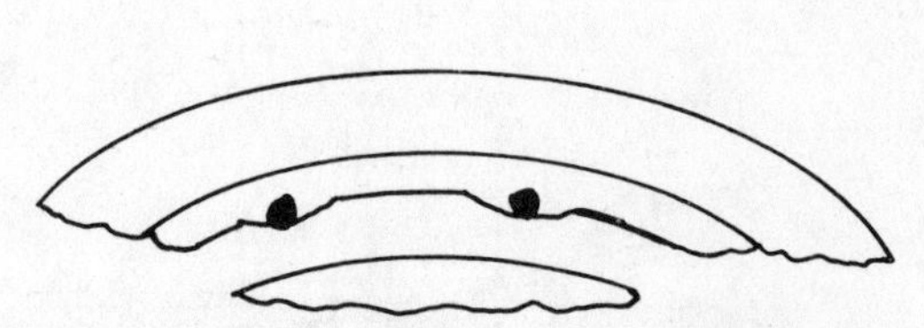

Fig. 6-32. Babbit remains the choice for light duty applications, in part because of its excellent embeddability. Smaller dirt particles are absorbed into the bearing for only a small reduction in running clearance (A). But larger solids are only partially embedded and score the journal with each revolution. (Courtesy Federal/Mogul Corp.)

Overlay Missing: A tapered crankpin or main journal will wipe the overlay from all or one side of each of the bearing halves. A bent rod will also remove part of the overlay, but on opposite sides of each bearing half. An out-of-round journal concentrates the wear near the bearing parting line. Pronounced wear on both halves of the center main bearings usually means a bent crankshaft or main bearing bore misalignment.

Radius Ride: This condition, characterized by wear on the edges of the bearing, is caused by an improperly ground crankshaft. The cure is to regrind the crankshaft and replace the affected bearings.

Fatigue Failure: Craters form in the bearing surface as the metal flakes off. Drag racing usually fatigues the upper rod bearings; high speed operation takes out the lower rod bearing shells. Main bearings might exhibit similar damage, but the location is different: acceleration pounds out the lower bearing halves, and high rpm fatigues the upper halves. Fatigue can also be caused by an out-of-round journal.

CAMSHAFT

Hydraulic lifters are normally replaced during a rebuild and must be replaced when a new camshaft is installed. Figure 6-33 helps to explain why: lifters and cam lobes are normally ground on a slight radius with 50 in. the norm for most American engines except Chrysler. (Early Buicks and Chevrolets did not have this feature.) The geometry rotates the lifters as they move, providing the wiping motion necessary to keep lifter bores clean and well lubricated. At the same time, the load is spread in a fairly wide band over the cam. A new cam, its tapered lobes in point contact with worn and flattened lifters, will be quickly destroyed.

A special tool should be used to extract lifters that will be reused; otherwise a pair of Vise-Grips works as well. Soak the lifter bores with carburetor cleaner and twist the lifters as you pull, "threading" them out of their bores.

Remove the camshaft, which is usually secured by a thrust plate behind the cam gear or sprocket. Some engines—including the Chevrolet small block V-8 and 396, Buick 401, Olds 425 and 455, and Chrysler Slant-Six—secure the camshaft with the distributor drive gear. When this provision fails, severe wear occurs at the point of camshaft contact on the block. Discuss possible repairs with your machinist.

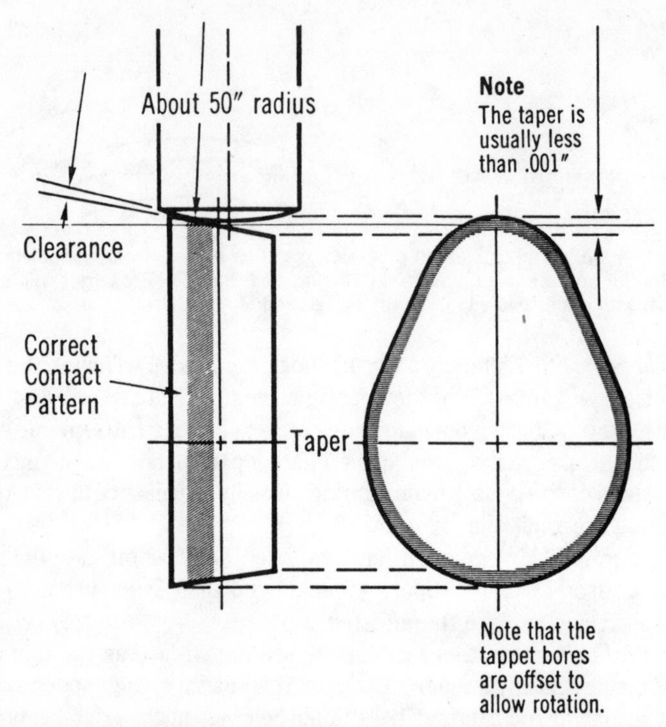

Fig. 6-33. A slight (about 0.002 in.) convexity on the lifter foot, coupled with a 0.0007- to 0.002-in. taper on the lobe and offset lifter bores, allows the lifter to rotate as it works. As the parts wear, the contact pattern shifts to the right. Eventually the lifter foot becomes concave and the wear track extends over the whole surface of the lobe. (Courtesy Sealed Power Corp.)

Carefully guide the camshaft out of its bore, avoiding contact between the fragile lobes and journal bearings. Withdrawal is easier if the block is raised to the vertical position.

The machinist will check camshaft straightness and will make the following measurements:

☐ Lobe minor axis diameter
☐ Lobe major axis diameter (Fig. 6-34)
☐ Journal diameter.

Ohv insert-type cam bearings are extracted and, once the block has been chemically cleaned, replaced with a special tool (Fig. 6-35). The operation is critical, in that bearing oil holes or grooves (e.g., Chevrolet small block) must be indexed with oiling passages in the bearing bosses.

BLOCK RECONDITIONING

Standard block prepping includes hot-tanking the block and other iron parts, replacing freeze plugs, hand-scrubbing all oil passages, and miking the cylinder bores throughout their lengths (Fig. 6-36).

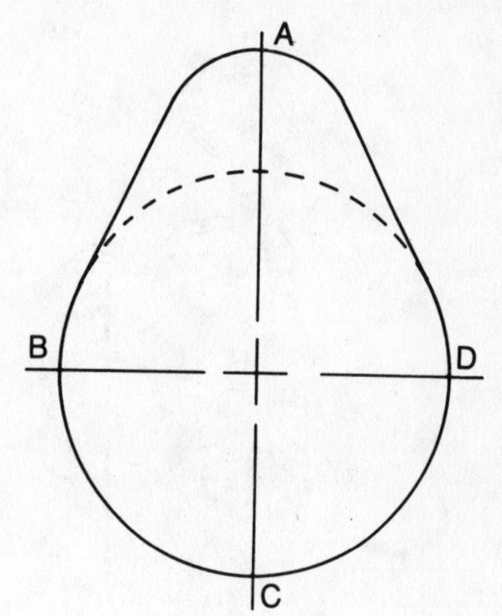

Fig. 6-34. Most camshaft wear occurs on the larger diameter to reduce valve lift, which explains the difference between AC and BD. (Courtesy Navstar)

Cylinder Bores

Some European engines, including certain Fiat, Peugeot, and Renault models, together with a number of industrial diesel plants adapted for automotive service, employ replacable cylinder liners, or sleeves.

Construction

Most of these sleeves are "wet," which means that sleeve OD is exposed to coolant and the ends are sealed with gaskets. Wet liners are a mechanic's dream, since liners, seals, and pistons are replaced as an assembly: any machining is limited to gasket mating surfaces, which may be subject to corrosion damage.

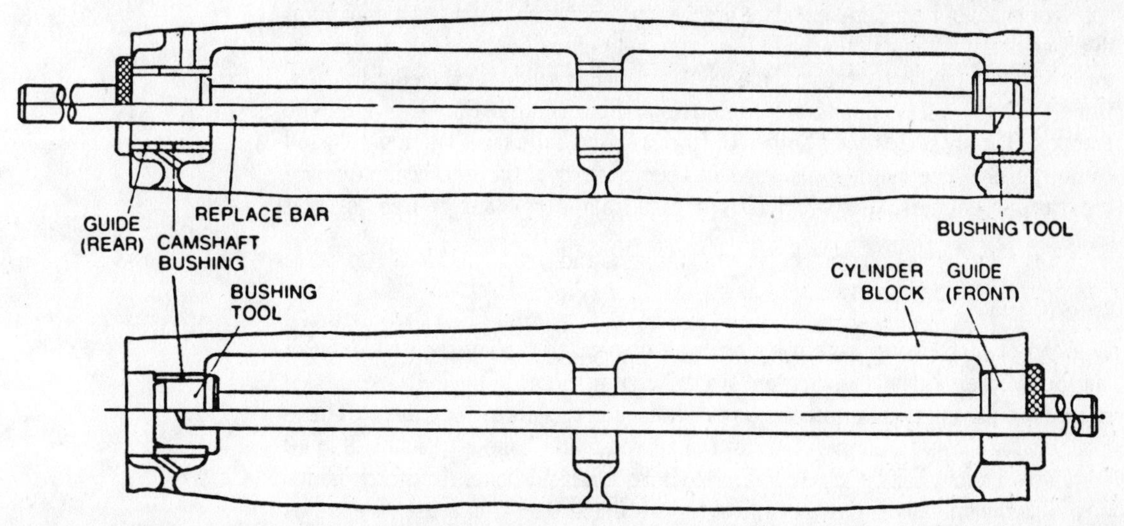

Fig. 6-35. Cam bearing service requires a special tool and care to see that the oiling circuit remains open. (Courtesy Chrysler-Nissan)

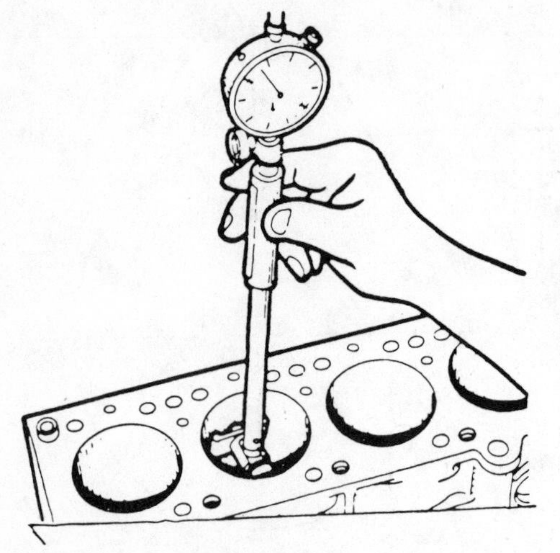

Fig. 6-36. Professionals use a cylinder bore gauge to determine wear, taper, and out-of-round. The same measurements can be made with a snap gauge. (Courtesy Chrysler Corp.)

Although air-cooled, early VW engines use a similar construction, in that piston/cylinder assemblies are replaced as matched assemblies.

Dry liners press into counterbores in the block, eliminating gasket problems, but at the cost of a slight thermal barrier between the bore and coolant. These liners are thick enough to allow one or two overbores and, when the bore limit is reached, can be replaced. Figure 6-37 illustrates a liner jack.

The integral cylinder bores on all domestic and most foreign engines are fodder for the boring bar.

Prestressing

As indicated earlier, cast iron is elastic and must be stressed to some approximation of its assembled state before machining can begin. As a minimum, the machinist will install the main bearing caps before boring the cylinders. A torque plate should be bolted to the head to simulate head bolt-induced distortion. But few shops will be able to accommodate any but the most popular domestic engines. Torque plates are a bother to fabricate and expensive when purchased commercially. The average shop would need dozens of them to accommodate the various engines, few of which share the same cylinder centers or bolt patterns.

Machining

The common procedure is to use a boring bar to make the initial cuts and finish out the cylinder with a hone chucked in a drill motor. Some of the better-equipped shops depend upon an automatic power hone for the entire operation. These machines are fast—a Sunnen CV can resize and finish-hone a typical V-8 in 40 minutes, or about half the time required to bore and hand-hone. Automatic hones give better dimensional control and a superb finish that is at least as good as OEM.

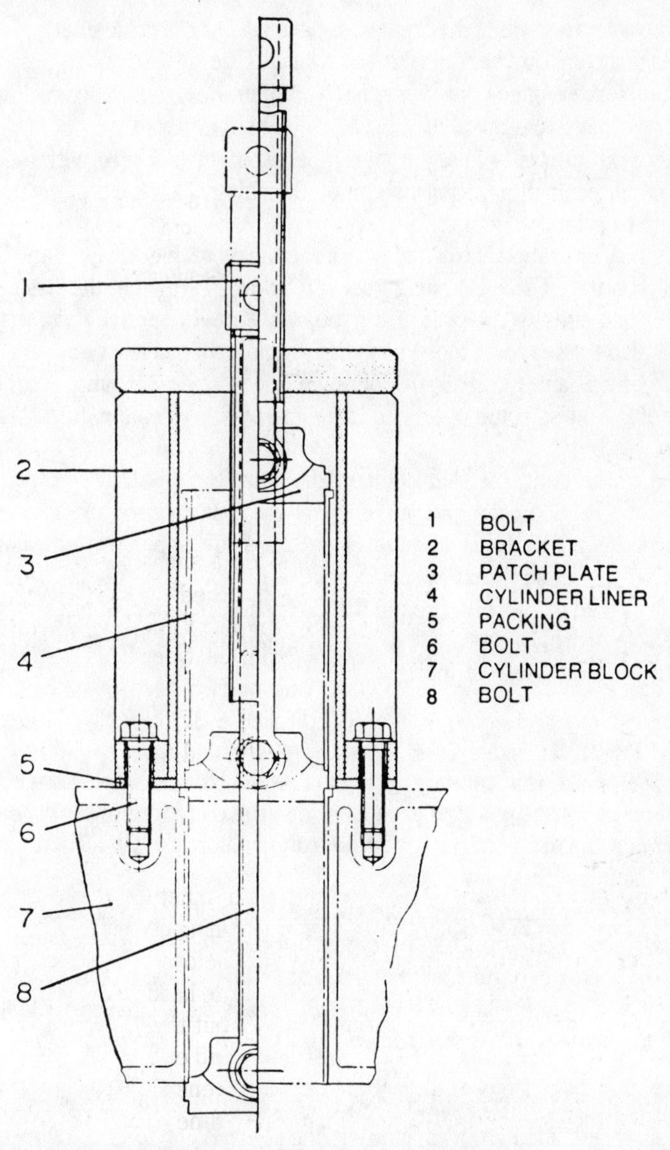

Fig. 6-37. Diesel tooling often includes a liner jack, as shown for Chrysler-Nissan.

The accuracy of the job can only be as good as the datum, or the reference point from which all measurements are taken. Most boring tools index to the deck, which is okay if the deck is parallel to the crankshaft centerline. Better results can be had if the cutter indexes to the main bearing saddles. Cutting begins by centering the tool on the cylinder with most wear. The overbore required to clean up this cylinder determines the size for those remaining. "Factory rebuilders" normally take out 0.030 in. because of the easy availability of pistons. Premium rebuilders are more considerate of your block and try to hold the overbore to the minimum.

Allowable oversize is determined by availability of replacement pistons and, to a lesser extent, by head gaskets. Available metal sets the upper limit. You can't take some engines out more than 0.030 in. before compromising cylinder integrity; others tolerate 0.120 in. without striking water.

The next step is to chamfer the upper ends of the bore with a 60-degree cutter mounted in the boring machine. Breaking the sharp edge makes ring installation easier.

The final operation is to hand-scrub the bores with detergent and hot water as detailed in the "Glaze Breaking" section below. Verify that this has been done.

Reboring adds to the cost of the job, since new oversized pistons must be used. It also makes for a more powerful engine, because of both the increase in displacement and the marginal increase in compression ratio. Fuel economy will fall off a bit and some overbored engines may be "cold natured" and balky on start-up.

One of the concerns that engine rebuilders have is that electronic engine controls might someday put an end to the practice of cylinder reboring. Displacement is the basic parameter, and a truly sophisticated management system would sense if that parameter were out of spec. To date, no one has, with any degree of success, broken the algorithm, the coda of rules that controls how vehicle microprocessors respond to sensor inputs. Future systems are not going to be any easier to fool.

Glaze Breaking

Cast iron is metal that can be run against itself. Iron pistons survive very well in iron bores. This is possible because the surface layer of the iron compresses and glazes to form a hard porcelain-like rubbing surface. Small scratches in the glaze fill and, as they say, "heal."

On the other hand, a mirrorlike surface does nothing to promote ring break-in. New rings need "tooth" to seat and for oil adhesion.

When honing is performed on standard-sized cylinders, that is, on cylinders that have not seen the boring bar, the process is called glaze breaking. If your budget will allow it, have the machinist do the work, particularly if he uses an automatic hone.

When viewed under a microscope, a correctly honed bore consists of a series of sharply defined grooves, intersecting at angles of 30 to 45 degrees relative to the bore centerline. Diamond-shaped raised sections of the bore, known as plateaus, account for 50 to 70 percent of total area (Fig. 6-38). The grooves

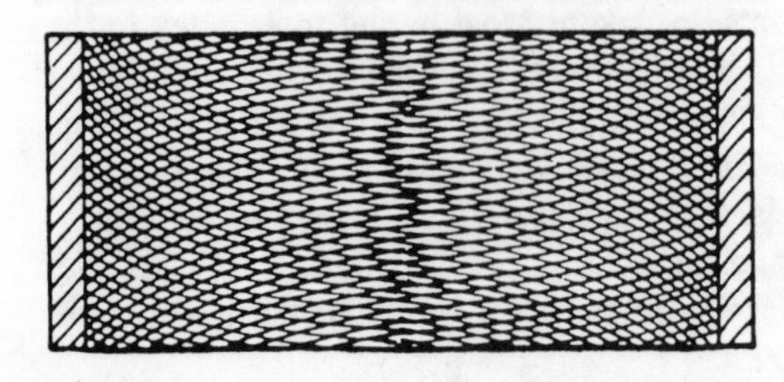

Fig. 6-38. A properly honed cylinder looks like this, with sharply defined grooves, diamond-shaped plateaus, and no evidence of metal folding or tearing.

act as oil reservoirs and as havens for metal particles. These requirements are easy to meet with automatic honing machinery. But a drill press or heavy-duty, low-rpm drill motor can do satisfactory work.

Either a brush hone, such as a 120-grit BRM Flex Hone (Fig. 6-39), or a more versatile spring-loaded hone may be used. The latter accommodates different cylinder diameters and features replaceable stoves. Precision hones, which are set to diameter by hand, are designed for serious metal removal and should not be used for glaze breaking.

The relationship between stone grit, rotary, and reciprocating speed varies according to cylinder diameter and ring type. General recommendations follow:

Bore Diameter (inches)	*Spindle Speed* (rpm)	*Reciprocations* (strokes/min.)
2	380	140
3	260	85
4	190	70

Grit selection is guides ring material: in general, chrome rings need a rough finish, such as that obtained with a 280-grit stone. Moly-filled rings seat with something finer, on the order of 400 grit. But recommendations vary among manufacturers.

Flood the hone with approved lubricant, such as mineral oil or PE-12, which is manufactured by AGS and available in aerosol cans. If using a spring-loaded hone with a drill press, set the stroke limit so that the hone is captive at both ends of the bore. A free-running hone can fling its stones. When using a hand-

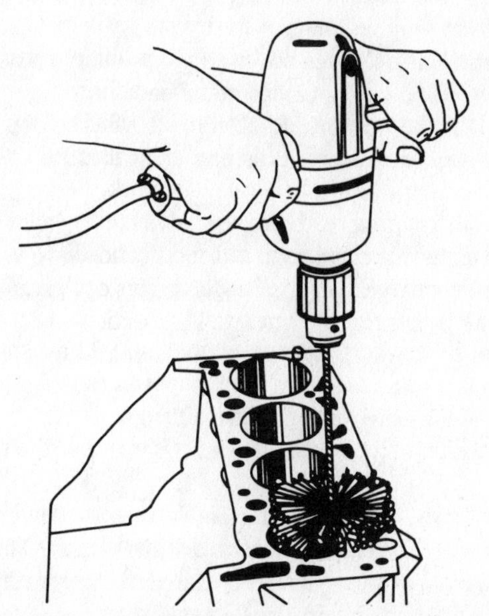

Fig. 6-39. Most mechanics break glaze with a brush, or ball, hone.

held drill, mark the lower limit of travel with a piece of masking tape wrapped around the hone shaft. These precautions do not apply to brush hones.

Keep the hone moving—whenever it is turning it is cutting—and do not pause at the ends of the strokes. Hone for three or four seconds and stop to inspect the bore. If the crosshatch pattern is flat, speed up the reciprocation or slow the spindle speed. 15 to 20 seconds of honing should be enough to do the job. Withdraw the bush hone on the final stroke while it is still turning.

Few mechanics clean the bores properly. Never use solvent, which merely floats the abrasives left by the hone deep into metal. Detergent, hot water, and a scrub brush are the prescription. Scrub the bores until the suds are white. Dry with paper shop towels, and repeat the process until the towels no longer discolor. According to TRW, half of all ring failures during break-in are the result of improperly cleaned cylinder bores. Oil the bores and deck immediately to prevent rusting.

A New Lease on Life

Engines with integral cylinder bores can be dry-sleeved to OEM specifications. But the process is expensive—about $50 per hole—and rarely used as an antidote for bore wear. A block so worn that it cannot be overbored is best replaced. Sleeving is a way of repairing damage to a single cylinder, such as that caused by detonation-induced cracks, a wandering wrist pin, or a broken valve. Damage must be near the center of piston travel and should not extend to the top or bottom of the bore.

Even so, it is nice to know that any engine can be sleeved. It gives the touch of eternity to a very ephemeral object. Racers sleeve their engines down to qualify for displacement classes or to gain additional block rigidity. Collectors sometimes resurrect vintage engines in this way.

One-eighth-inch-thick sleeves are used when a section of the cylinder wall has broken out and when the block has metal enough to leave 0.062 in. of iron between sleeve OD and coolant. Replacement sleeves are dry sleeves and require solid foundations. Local cracks and scores call for 3/32-in. (0.093-in.) sleeves.

The aftermarket supplies ready-made sleeves for popular domestic engines and uncut sleeves in various diameters. Molded sleeves have almost disappeared; most sleeves are centrifugally cast in nodular iron and given a better wearing surface than OEM blocks.

The machinist bores out the block to give a 0.003-in. interference fit with sleeve OD. Honing is not necessary. Some operators make a plunge cut along the whole length of the bore; others leave a flange, or lip, at the bottom of the bore to create a positive stop for the sleeve. Otherwise, the sleeve might drop when bored.

The machinist may use a lubricant, such as white lead or red lead oxide and glycerin on sleeve OD. Others prefer a dry installation. Normally, the sleeve is pressed into place and then ground or flat-filed flush with the deck. A few shops cool the sleeve (a sub-zero freezer will shrink sleeve OD about 0.001 in.) and heat the block. If the machinist works quickly, he can drop the sleeve home before temperatures equalize. The sleeve is then bored and honed to piston size.

Align Boring

Align boring or honing is a fairly exotic operation that is not often needed or performed. The job is usually occasioned by a spun main bearing that has chewed its cap, distorted caps that have lost register with the block, or a warped block. If the machinist can handle the work, he will determine alignment with a mandrel, then shave 0.005 in. off the cap/saddle mating surfaces. Caps are assembled, torqued to spec, and resized with a boring bar or hone (Fig. 6-40). The less metal taken off during the parting process, the better. Align boring introduces slop into the timing chain by displacing the crankshaft upward, toward the cam. The effect is to retard valve timing.

Although we're getting ahead of ourselves, alignment can be verified during assembly. Install new main bearings, lubricate them as described in the next chapter, and lay the crankshaft. Install the caps—without the rear seal—and torque to spec. Spin the crankshaft—if it binds you have a problem. Either the block is warped or the machinist did not check crank straightness.

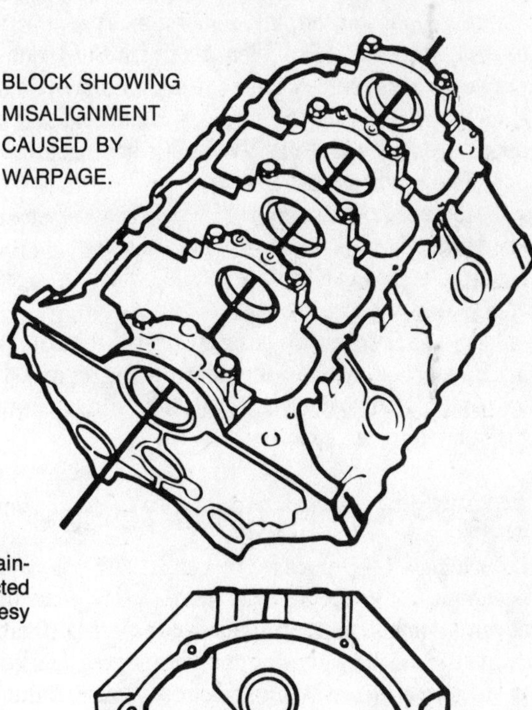

Fig. 6-40. Block warp and main-bearing cap stretch can be corrected by precision honing. (Courtesy Sunnen Products Co.)

Decking

Decking is one of those terms like "blueprinting" that mean whatever one wants them to. In the purest sense, decking means to mill the block fire deck square with the crankshaft, so that tdc can occur at the same point in each cylinder. That, plus a dead accurate check of the crankshaft and selective assembly of pistons and rods, will give all cylinders the same compression ratio.

In the rebuilder's sense, decking means to mill or grind a warped block reasonably true and square. The operation is rarely necessary and relatively expensive. Any serious amount of decking leads to complications with the intake manifold on vee engines (see Chapter 5).

Balancing

You may wish to investigate the possibility of balancing the engine. The procedure may add several days to the turnaround time—most shops farm the work out—and will add $75 to $150 to the cost. A statically and dynamically balanced engine will vibrate less and be perceived as accelerating with less effort. However, balancing is no panacea.

Engine balance is a function of geometry. In-line four-cylinder engines are inherently imbalanced, and nothing short of a Mitsubishi-type Silent Shaft can do much about it. (The Silent Shaft runs at twice crankshaft speed to dampen primary and secondary forces. And, truth be told, owners may not notice when a rebuilder discards the trouble-plagued shaft.) In-line sixes and horizontally opposed engines, such as VW Beetle and Subaru, have a geometrical advantage over in-line fours and all vees.

Another limitation is in the balancing process. According to the formulas used by makers of the equipment, engine balance occurs over a fairly narrow rpm range. You can balance at idle, which is impressive, and be out of balance at highway speeds.

And as Smokey Yunick points out, the rotating parts of an engine spin in a vortex of hot oil, like cotton candy clinging to a stick. Oil mass and distribution vary with load and rpm. It is difficult to understand how balance could be maintained under such conditions.

THE SHAPE OF THINGS TO COME

Rebuilding street engines is a conservative activity, governed by tradition and economics. Except for the more widespread use of aluminum, plus some advances in lubrication and gasket materials, the technology remains as it was in the 1950s. However, some interesting things are happening in the area of protective coatings. We have already mentioned TRW Plastic Ceramic Piston Rings.

Swain Tech TBC uses an analogous process to coat almost all internal engine and transmission parts. The company's Thermal Barrier Coating (TBC) is a permanently bonded cerametallic coating that reduces heat transfer when applied to piston tops, combustion chambers, valves, and turbocharger exhaust manifolds. Swain Tech has applied similar coatings to critical parts on the Space Shuttle.

Less heat transfer to the cooling system translates as more power to the crankshaft. A by-product is cooler-running pistons that, because they are shielded, maintain their mechanical strength at elevated chamber temperatures and require less bore clearance.

According to the company, the most dramatic demonstration of the effectiveness of this coating occurred at Lime Rock, Connecticut, when Paul Newman's turbocharged Nissan experienced sudden loss of coolant. Temperature gauge readings remained normal, and Newman was able to continue for two laps at full power before the uncoated cylinder head burned through. Upon teardown, the TBC-treated pistons showed no appreciable wear and could have been reused.

Swain Tech also markets a polymer moly-tungsten-disulfide coating for friction reduction. It may be applied to piston skirts, timing chains, and sprockets, and to transmission parts for a claimed increase in horsepower, durability, and fuel mileage.

Further information can be obtained from:

Swain Tech Coatings
Racing Division
35 Main Street
Scottsville, New York 14548
(716) 889-2786

7
CHAPTER

Engine Assembly

AT THIS POINT, THE ENGINE IS FRAGMENTED INTO A HUNDRED OR MORE PIECES, some of which elude recognition, and scores of new parts are stacked about in their cartons. This collection of pieces, plus a completed cylinder head, a crankshaft, a block verging on rust, and a stack of crumpled receipts is all you have to show for your trouble. But the potential is there for an engine that will be nearly as durable, at least as powerful, and much more satisfying to own than it was on the day it left the factory.

Assembly always begins with the crankshaft, followed by piston and rod installation. At this point, engine architecture determines the sequence: on ohv engines the camshaft and cam drive is usually next, followed by the front cover, harmonic balancer, flywheel, oil pump, oil pan, and head. The cylinder head must be installed early on ohc engines in order to establish valve timing. Once this is done, the front cover, harmonic balance, and flywheel can be installed. The oil pan (which almost always attaches to the front cover) is one of the last components to be bolted up.

THINGS TO REMEMBER

1. A book can't cover all possible mistakes in engine building. Most assembly errors occur in the following areas:
   - ☐ Bearing fits
   - ☐ Gasket compatibility
   - ☐ Overlooked fasteners

Take the time to satisfy yourself that everything is right. This is the last free look you will have. When you see engine internals again, it will be because something went wrong.

2. Read the factory manuals. The days when an engine could be "cold-turkeyed" together on the basis of instinct and an elbow torquometer are long gone.
3. Work clean. One acquaintance assembled a small-block Chevrolet in his living room, which may be taking things a bit far, but nobody could fault the intention.
4. Go heavy on the oil, lubricating everything except the backs of insert bearings and gasket surfaces. Proprietary assembly lubes are available for bearing and O-ring lubrication, but most mechanics still use motor oil. API-SE, 30-weight motor oil, should give adequate start-up protection, although you may wish to mix the oil with an approved friction-reducing additive, that carries a GM, Ford, or Chrysler part number. Head bolts and other torque fasteners require straight, undoctored motor oil. As nearly as can be determined, factory torque specs are written for oil-lubricated fasteners. Antifriction additives or specialized assembly greases will fool the wrench and result in over-torquing. EP (extreme-pressure) grease is mandated for new camshaft/lifter combinations and should be used on all of them. Any assembly lube that is labeled as suitable for camshaft installation is suitable. Crane and other cam grinders market proprietary products for this purpose.
5. Use new gaskets and, where recommended, sealant. Normally, sealant is applied to composition (paper-like) gaskets used in the cooling system. Sealant is not to be applied to rubber gaskets or O-rings.
6. Work slowly. Time is on the side of the amateur mechanic. Decipher your notes and become comfortable with the factory manual. One cutaway drawing is worth 10,000 words.
7. And finally, do not hesitate to consult with your machinist. Technical support is part of his professional responsibility.

CRANKSHAFT

Install the upper bearing halves in the main bearing webs, noting that all bearing halves may not be interchangeable. Flanged thrust bearings are impossible to mistake for others, but there also may be more subtle differences. At one time it was common for both shells in a main bearing set to be identical. That is, both were drilled and grooved for oil. Current practice is to increase available bearing area by confining the drilling and grooving to the web (or engine) shells. No oil enters from the main bearing caps. Interchanging the two bearing shells will destroy the engine.

In the 1960s, American manufacturers experimented with bearing clearances and sometimes specified more clearance on No. 1 main than on the others. This was supposed to increase bearing life by allowing the crankshaft to flex under the side loads generated by belt-driven accessories.

The front upper bearing on the 2.3-liter Ford engine, illustrated in Fig. 7-1, is distinct from the others and carries its own part number. This engine is also

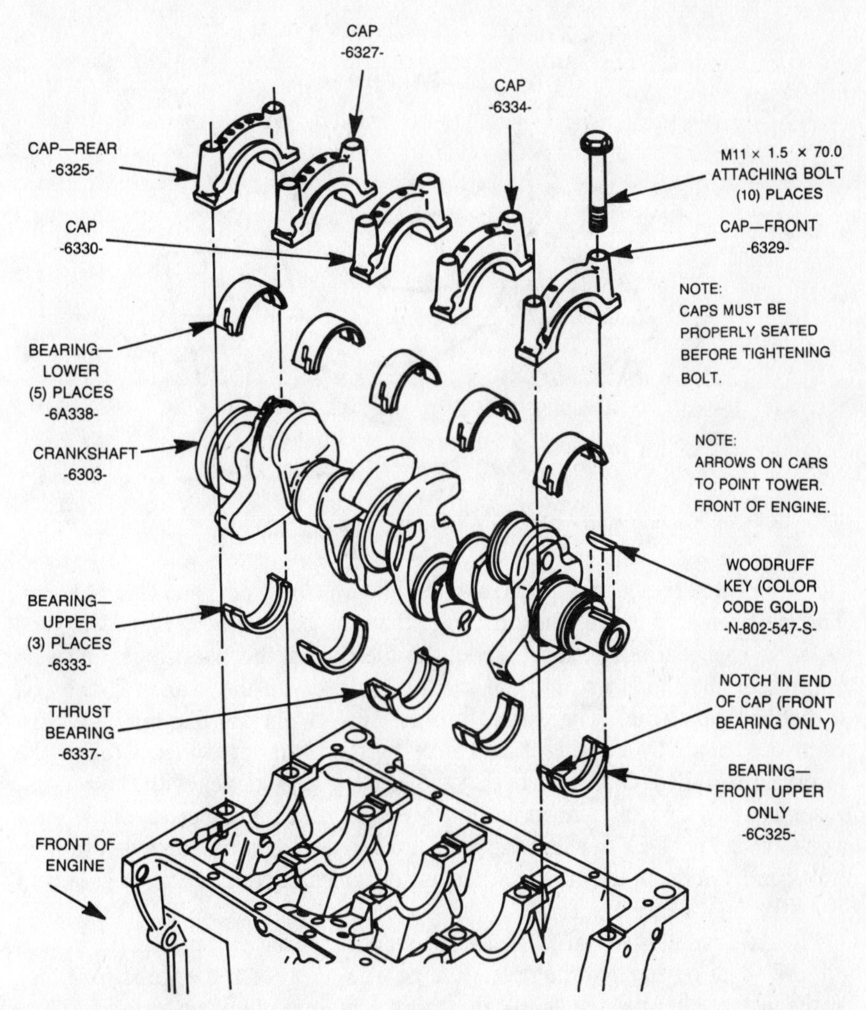

Fig. 7-1. Not all bearing shells in a typical main bearing set are identical. The 2.3-liter Ford illustrated here uses a special front upper bearing, and an unusual thrust bearing that, at first glance, might appear to be a misbuild. Only the center upper bearing is flanged for thrust; the lower bearing is identical to the other lower bearings. Industry practice is to flange both upper and lower center bearings. (Courtesy Ford Motor Co.)

unusual, in that only the upper center main bearing is flanged to accept thrust loads.

Note: Thrust bearing flanges must make solid contact with the sides of the upper main bearing saddle and with the sides of the bearing cap. Otherwise, thrust bearing clearance will be insufficient. A few bumps with a rubber mallet will seat the flanges of the installed bearing shells into firm contact.

Remove a main bearing insert set from its protective wrappers and verify that the size stamped on the OD of each bearing is what has been ordered. Note the presence of the locating tab, which, upon assembly, must index to recesses in the saddle and cap (Fig. 7-2).

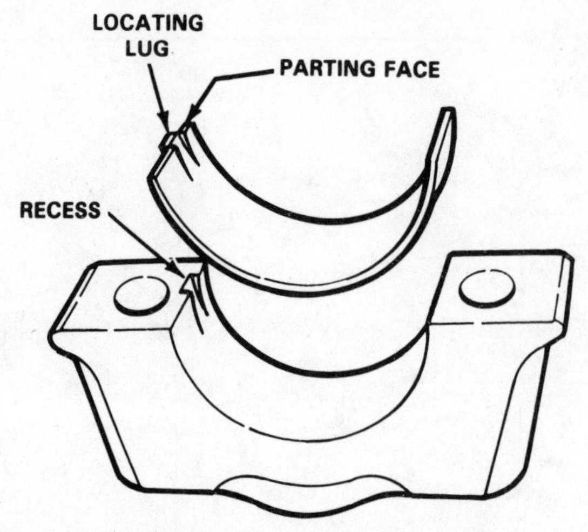

Fig. 7-2. Insert bearing nomenclature. (Courtesy Federal-Mogul)

Check that the bearing saddles, or webs, are clean, dry, and free of burrs. The free diameter of the insert is larger than the bearing bore (Fig. 7-3), and as far as main bearings are concerned, this means that the insert must be rolled into place. Slip the insert into the saddle, leading with the plain end (the end without a locating tab). The partially installed insert will look like Fig. 7-4. Now complete the job by applying thumb force to the insert parting face, rolling the insert home. Make dead certain that the locating tab indexes with the recess. When installed correctly, insert parting faces will stand equally proud of the web, as shown in Fig. 7-5. This is a design feature to assure retention when the cap is torqued. Lubricate the insert with motor oil, smearing the oil with your fingers over the total bearing surface.

Install bearing shells on the remaining saddles, making certain that the inserts are drilled for oil and that the drillings align with oil ports in the webs. A glance at the installed bearing shells will show this, but some shops go a step further. The mechanic must physically verify that oil ports are open by running a drill bit or a length of welding rod through the bearing shell and into the web—just to make sure.

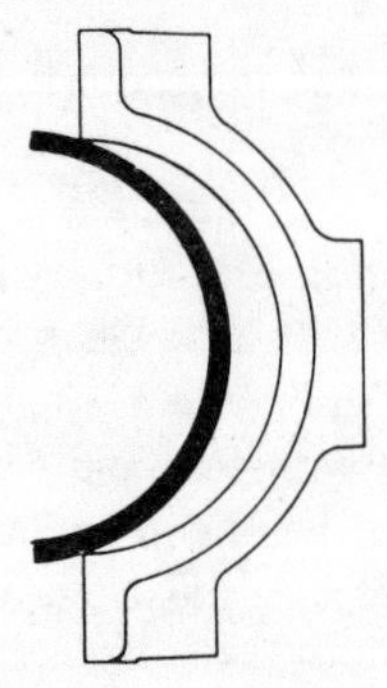

Fig. 7-3. Spread, representing residual tension between insert OD and bearing boss ID, assures that the bearing will seat around its full diameter. (Courtesy Sunnen Products Co.)

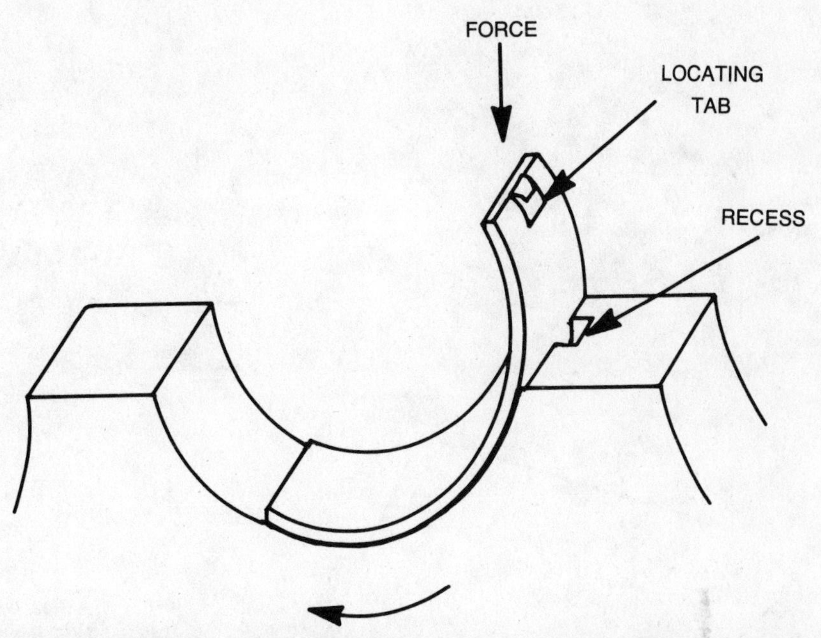

Fig. 7-4. Main bearing inserts roll into place, exactly as if you were scooping ice cream. Slip the bearing into the cap or saddle, leading with the smooth, or untabbed, edge. Then, with your thumbs on the parting face, push the insert home, seating the locating tab in its matching recess.

You may wish to pre-assemble the caps with bearing shells installed and measure bearing ID for comparison against crankshaft OD. Most main and crankpin bearings are slightly ovoid, with the larger dimension at the bearing parting faces. This is a design feature and nothing to be concerned about, so long as the running clearance is within the specified range at all points on the bearing circumference.

Wipe off any dust that might have accumulated on the crankshaft, and oil the journals with the same thoroughness as the bearings. Align the crankshaft over the thrust bearing, and carefully lower the crank into place on the saddles. Install a bearing insert in the center main bearing cap, lubricating it as before. Repeat the process for the remaining caps.

Note: Cap number sequence must be correct and cap arrows or other directional referents must be aligned as originally found.

Rear Oil Seal

Most American engines still use a two-piece seal, integral with the rearmost main bearing. Either a rope or a rubber sealing element will be supplied in the gasket set. In the absence of specific instructions to the contrary, rope-type seals are installed as follows:

1. Roll the lower seal half into its groove with a large socket wrench or length of pipe (Fig. 7-5A).

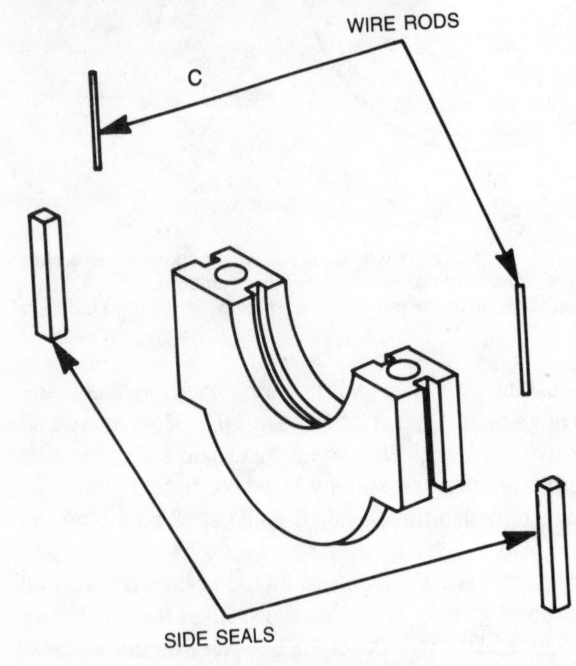

Fig. 7-5. A rope, or wick, seal should be rolled into place with the help of a large socket wrench or similar device (A). Once the rope is seated, the ends are trimmed flush with a razor blade (B). Only then is the seal lubricated with chassis grease. The rear main bearing cap may employ side seals, which should be installed as per instructions with the gasket kit. (Courtesy Fel-Pro Inc.)

2. Lightly lubricate the seal with chassis grease.
3. Trim the seal ends flush with a razor blade (Fig. 7-5B).
4. Repeat steps 1 through 3 for the block-side seal.
5. Some engines with deep block rails use side seals, as shown in Fig. 7-5C. The installation technique varies with the type of side seal supplied: rubber seals are sized to fit; fiber seals swell to size when wetted with oil.
6. The cap-to-block interface can be made oil-tight with an anaerobic sealant, such as Fel-Pro Set and Seal.

One- or two-piece rubber seals are used on newer engines and sometimes can be retrofitted to earlier models. For example, GM has developed a one-piece rear seal for its 173 V-6 engine, which is available under PN 14081761. The seal is packaged with a throwaway installation tool. Some 1985 production GM 122 and 173 engines use a one-piece seal that can be substituted for the two-piece seal on earlier units. The revised part is available from McCord as PN B5301. Rubber seals should be lubricated with motor oil, and double-lip oil seals should be filled with moly grease to prevent damage during initial start-up.

One-piece rear seals, which mount between the block and flywheel, are installed after the main bearing caps are torqued. Thoroughly lubricate seal lips with motor oil.

Note: Lipped seals have one feature in common: the steep side of the lip goes toward oil pressure.

Cap Installation

Lightly oil the main bearing cap bolts, and using a torque wrench, tighten the caps. Begin with the center bearing cap, taking it down to one-third of final torque. And, moving in both directions away from center, tighten each of the caps to the prescribed one-third torque. The idea is to restrain the crank in progressive steps from the center outward, leaving the ends free to move until last. Repeat the sequence twice—tightening the caps to two-thirds torque and finally to full torque.

The crankshaft might drag on the seal (especially on a rope seal), but it should turn by hand. If it doesn't, recheck cap numbering and orientation. A cap might be out of sequence, reversed, or improperly aligned. Relax torque on one cap at a time until the offending cap is isolated. If the cap appears to be installed correctly, the problem may be misalignment that can sometimes be corrected by driving the cap to seat with a soft-faced mallet. Retorque and test.

If that doesn't solve the problem, remove the caps and inspect the bearings. Bright spots verify metal-to-metal contact. Remove both halves of any suspect bearing. Check bearing size, which will be stamped on the back of each insert, and look for burrs or dirt that would prevent the bearing from seating. Determine bearing clearance.

At this point, you have fairly well eliminated the possibility of assembly error; the fault is either in the crankshaft itself or with the block. Fortunately, block distortion of the kind that affects the lay of the crank is rare and is usually detected during disassembly, upon inspection of the main bearing inserts. The same goes for a bent crank. But if the crank has since been reground or straightened, it is almost surely at fault. Take the partially assembled block back to the machinist, and when he has done whatever corrective work is necessary, recheck main bearing clearances.

Plastigage

Plastigage is a very inexpensive and accurate method of verifying bearing clearance. You will want to purchase a package of green wire, sized for the narrow clearances expected in a new engine. The product is discussed here in the context of main bearings, but it is good practice to use Plastigage on every two-piece bearing. It *must* be used when you install selective fit bearings (e.g., the coded, variable-sized bearings used by Honda and, on some models, by Toyota).

Plastigage is a soft plastic wire inserted between the bearing and journal. When the cap is torqued, the wire flattens. A scale on the package converts wire width into inch-standard and metric bearing clearance.

To use Plastigage:

1. Wipe the oil off the journal and cap.
2. Lay a piece of Plastigage along the length of the journal.
3. Without turning the crankshaft, tighten the cap to the prescribed torque value.
4. Remove the cap and "read" the width of the Plastigage on the scale provided (Fig. 7-6). Compare this reading with the OEM spec, which should usually be in the neighborhood of 0.0015 in. Wipe off the Plastigage; the product will dissolve in oil, but it is not a good idea to have plastic floating around the engine. Do not forget to lubricate the bearing and crank journal before final assembly.

If the clearance is not within specification, take the block, crankshaft, main bearing caps, and bolts back to the machinist. He has a problem—a bent crank, misaligned bearing saddles, or, what is more likely: radius ride (see the section on bearing maladies in the previous chapter).

Crankshaft endplay, or thrust-bearing clearance, is checked as shown in Fig. 7-7. Using a prybar, lever the crank forward, then back. Check the clearance between the thrust bearing and rubbing face with a feeler gauge.

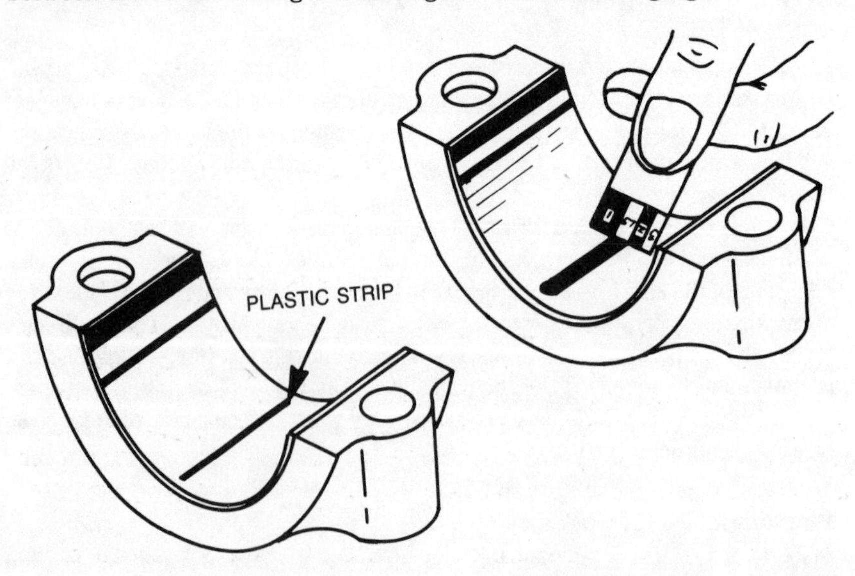

Fig. 7-6. Plastigage is an extremely accurate method of determining bearing clearance. (Courtesy Detroit Diesel Allison)

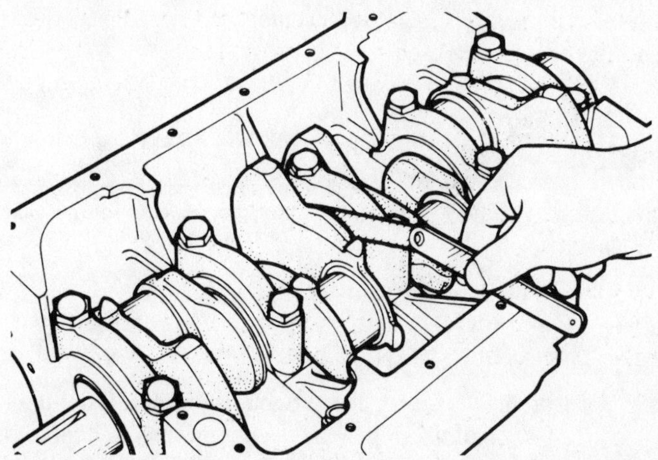

Fig. 7-7. Thrust bearing clearance is a critical parameter that must be checked. (Courtesy Chrysler Corp.)

Insufficient clearance is the usual problem. Normally you can correct this by bending the thrust bearing flanges, as described above, positioning the thrust washers properly in their grooves, or, as is sometimes the case, by aligning the thrust-bearing cap. This is done by loosening the cap, levering the crank hard forward, then backward. This seats the bearing. Tighten the cap while prying the crank forward.

If none of these stratagems works, or if the clearance is excessive, return the parts to the machinist.

PISTON ASSEMBLIES

The pistons are probably ready for installation now. The machinist will have assembled new pistons to the rods, together with the new wrist pins and pin locks that are part of the package. Used pistons will have been chemically cleaned, the ring grooves squared—and some machinists go so far as to install the rings.

Prepping

The alternative is to save a few dollars and prep the pistons yourself.

Piston tops. Piston cleanup begins by removing the carbon deposits on the crowns with a steel wire wheel. Do not wire-brush the piston sides: wipe off light accumulations of carbon above the top ring groove with a solvent-impregnated rag. Carbon on piston undersides and in the ring grooves is best attacked with a chemical solvent, such as Bendix Econo-Clean. Paint remover also works, although it is slow and some formulations can, with long exposure, leech trace elements from alloy.

Ring grooves. No chemical that is both safe to use and to dispose of can do more than soften the carbon buildup on the ring grooves. The carbon must be scraped off. Professional ring groove cleaners are miniature lathes that can

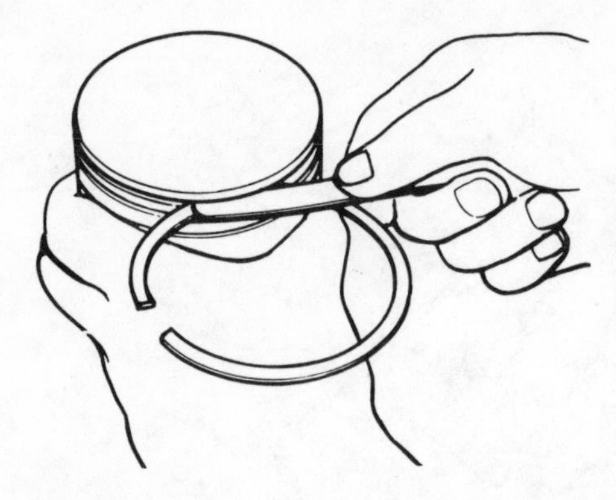

Fig. 7-8. Check ring groove width against factory specifications with a new ring and a feeler gauge. No. 1 ring groove is the one most likely to be hammered out.

be precisely set to remove carbon without gouging metal. Hand-held cleaning tools work, after a fashion, if the carbon has been previously softened. Otherwise the tool will butcher the ring grooves. It is almost as easy (and less risky) to use a discarded piston ring as a scraper. Break two or three inches off the ring, grind the jagged edge sharp and mount it into a file handle. The work will make you a believer in four-cylinder engines.

The grooves should be smooth and within spec as determined by measuring the vertical clearance with a new ring installed (Fig. 7-8).

Piston/rod assembly. When the original pistons will be reused, it is rarely necessary to separate them from their rods. New pistons must, of course, be mounted on the original rods. Before proceeding, determine (a) the leading edge of the piston, which is indicated by a mark on the piston crown as described below under "Rings"; (b) the orientation of the rod relative to some engine reference, such as the camshaft or oil pump. Do not proceed until you are certain of both of these referents, either from notes taken during disassembly or from the factory manual, which will always be explicit on these matters. Figure 7-9 illustrates piston/rod orientation for Ford small-block V-8s.

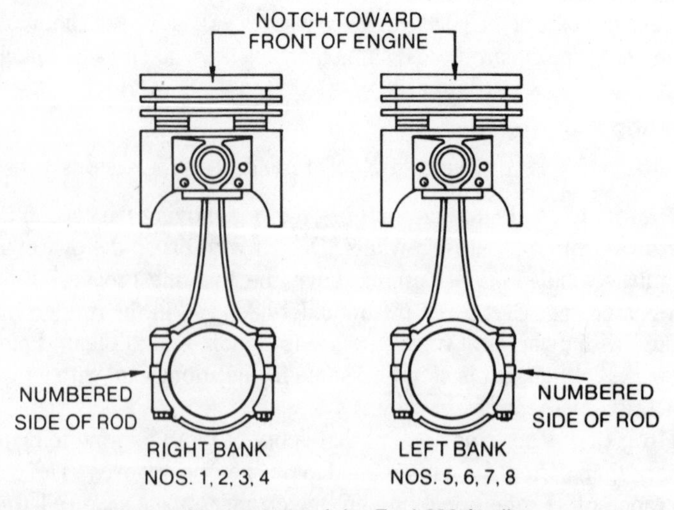

Fig. 7-9. Piston/rod orientation for members of the Ford 289 family.

Fig. 7-10. Spring locks, sometimes called circlips, secure the ends of the wrist pin. These locks should be renewed as part of the rebuilding process.

Floating piston pins, free to move on both the rod and piston, have become standard on modern engines. Spring locks prevent the pins from moving into contact with the cylinder walls (Fig. 7-10). But piston pins do not float at room temperature and are normally removed and installed with the aid of a press (Fig. 7-11). Heating the piston—either by rags soaked with hot oil or by placing the piston, crown down, on an electric hot plate—will also release the pin.

Bright spots on the connecting rod or piston bearing surfaces indicate excessive wear. As a rule of thumb, subject to correction by factory specs, pin clearance should be 0.0001-0.0003 in. at the piston and 0.0003-0.0005 in. at the rod. Increase rod clearance to 0.0005-0.0007 in. if the rod is drilled for full pressure lubrication.

These clearances will produce a light-to-moderate palm fit when the piston is heated and bearing surfaces are oiled. If the pin pauses or clicks at the second boss, the piston or rod is warped.

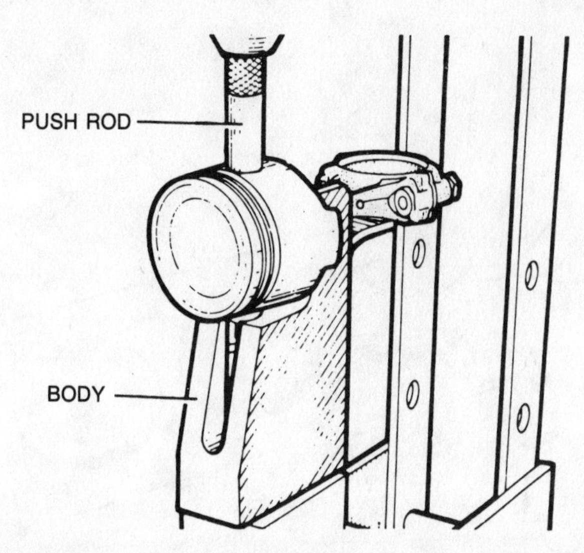

Fig. 7-11. Floating piston pins are adrift only when the engine is warm; when it is cold, the pins must be pressed out.

Always use new pin locks, compressing them no more than necessary to slip into the piston boss. Make absolutely certain that the locks seat in their grooves.

Rings

It might be up to you (since many shops do not do this) to place the correct ring, right side up, in the correct groove. Fortunately, the rings are packaged in individual ring sets with a detailed instruction sheet in each carton.

The forward, or leading, edge of the piston will be marked with a notch, arrow, or the letter "F," as shown back in Fig. 7-9. If it has not already been done, verify that connecting rod oil holes or numerical match marks are in correct orientation with the piston.

It is good practice to measure ring gaps—the distance between the ring ends—for the two compression rings and for one-piece oil rings. But this is not done if the machinist has already installed the rings on the pistons. In that case, the gap was his responsibility to measure, and in any event the stress of removing and installing the rings is more damaging than a slight variation in ring gap.

Figure 7-12 shows how the end gap measurement is made: the ring is squared in the bore with a piston and the gap compared to the ring manufacturer's specification. Check each one-piece ring of a set against the bore it will run in. Narrow gaps can be opened by passing the ring ends over a file mounted vertically in a vise.

The machinist should have checked ring-to-groove clearance (which is described above), but it doesn't hurt to take a look at it, particularly if the original pistons are going back in the engine.

Installing rings on the pistons requires a special tool, such as the one shown in Fig. 7-13. Attempting to mount the rings by hand is dangerous; ring edges

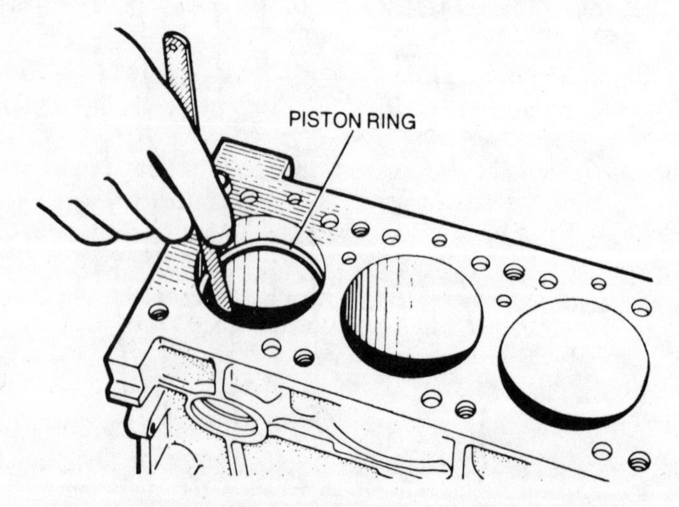

Fig. 7-12. Make the end gap measurement with the ring positioned about 2 in. below the deck as shown. A piston is used to square the ring in the bore. (Courtesy Chrysler Corp.)

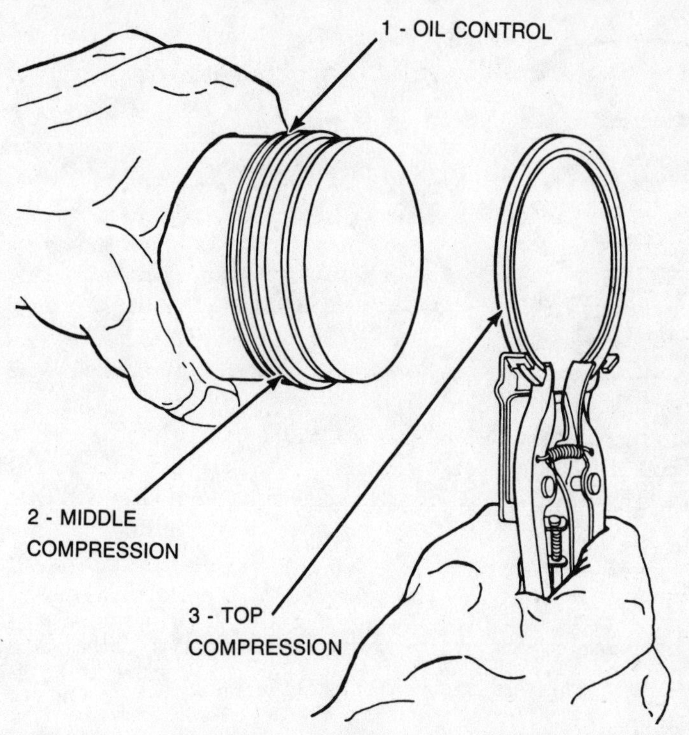

Fig. 7-13. Fingers cannot substitute for a professional-quality ring expander. (Courtesy Kohler of Kohler)

are deceptively sharp and will almost surely overstress the rings and lead to early failure. The oil ring—No. 3 counting from the top of the piston—is installed first (Fig. 7-14). If it is a one-piece ring, use the expander tool, opening the ring just far enough to slip it over the piston. Most oil rings do not have a definite up or down orientation, but check for reference marks on the ring sides.

Three-piece oil rings, the most common type, are installed by hand. The central, segmented part of the ring goes on first. Figure 7-15 shows the standard configuration, with ends butting. Next, roll on the bottom rail, which fits into a groove in the central section. The rail gap is usually 45 degrees away from the butted ends. Finally, install the top rail, with its gap offset 45 degrees in the opposite direction, as shown. Carefully inspect the assembly to verify that the segmented-section ends do not overlap and that the rails are captive in their grooves. A properly assembled ring will move under finger pressure.

No. 2 compression ring goes on next. This ring is almost always unidirectional. One side will be marked "T," "Top," or sometimes "O" (on European engines), and that side must be up. Expand the ring no more than necessary to squeeze it over the piston. Note the marks on No. 1 ring, which is also unidirectional, and install, again being careful not to overstress it in the process.

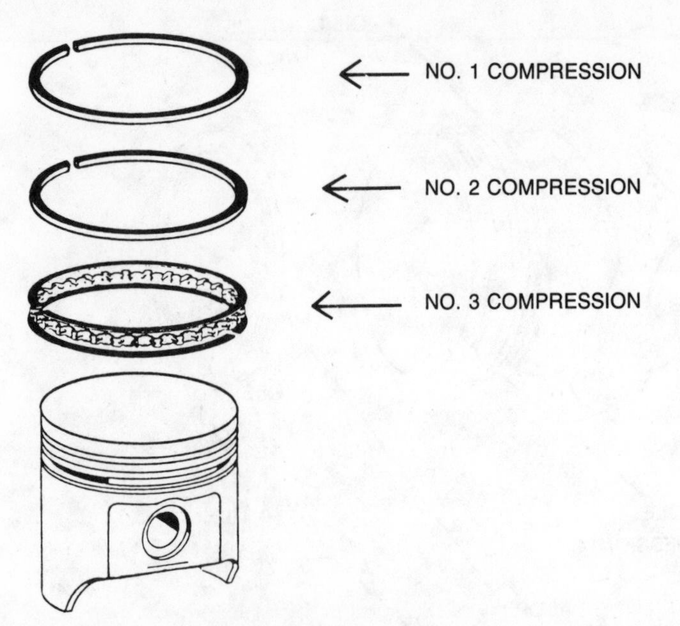

Fig. 7-14. Three-ring construction is the norm for automotive pistons. (Courtesy Sealed Power Corp.)

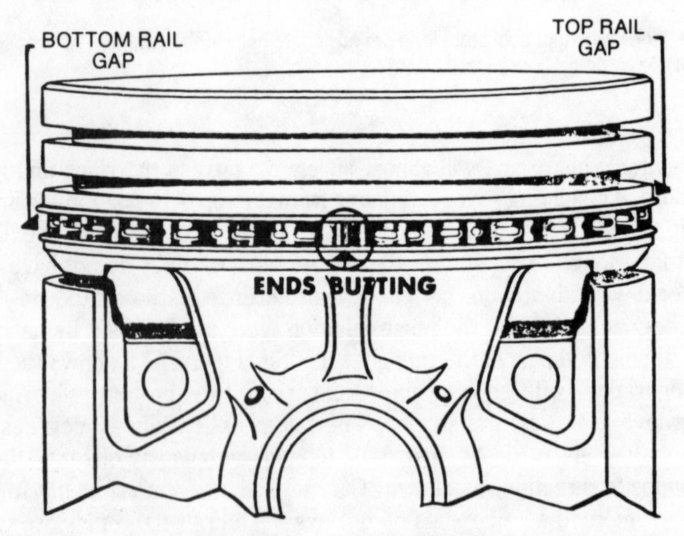

Fig. 7-15. The oil control ring is installed in three steps: central segmented element (ends butted as shown), lower rail, and upper rail. (Courtesy Sealed Power Corp.)

Lubricate the rings and piston pin. A sloppy but effective way to do this is to dip the piston in a coffee can half-filled with motor oil. Stagger the ring gaps in the pattern indicated by the ring or engine maker. The rings will not long remain in this orientation: they are free to rotate and, according to one study, reach a speed of about 80 rpm. In any event, one would not want to install the rings with their gaps aligned.

Piston Installation

Follow this procedure:

1. Remove the bearing cap from No. 1 rod and piston assembly.
2. Install an insert bearing into the rod, indexing the oil hole in the insert with the oil hole in the rod bore.
3. Lubricate the bearing, smearing motor oil over its entire surface. Insert the lower crankpin bearing into the cap, referencing oil holes (or their absence), and lubricating as before.
4. Slip lengths of fuel or vacuum-line hose over the ends of the connecting rod bolts, as shown in Fig. 7-16. The hoses act as bumpers to protect the crankshaft.
5. Rotate the crankshaft to bring No. 1 crankpin to bottom dead center. Lubricate the crankpin with motor oil, wetting the whole bearing surface.
6. Install the ring compressor over the piston, so that the lower edge of the tool is about ½ in. below the oil ring. Ideally, you should tighten the compressor bands just enough to compress the rings for entry into the bore. The illustration in Fig. 7-17 shows this being done; note that the mechanic pushes the piston down as he tightens the compressor. Overtightening will do nothing more than bind the tool to the piston, making extraction difficult.
7. If more squeeze is needed, because the compressor is either poorly designed or distorted, it will be necessary to tap the piston out of the tool with a hammer handle. Go gently and stop if the piston hangs.

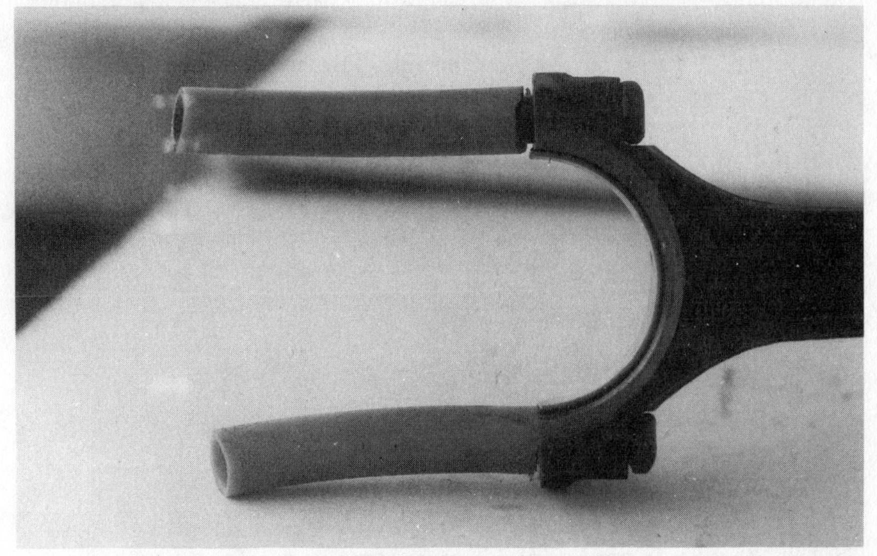

Fig. 7-16. Fuel line over the rod bolts prevents crankshaft damage.

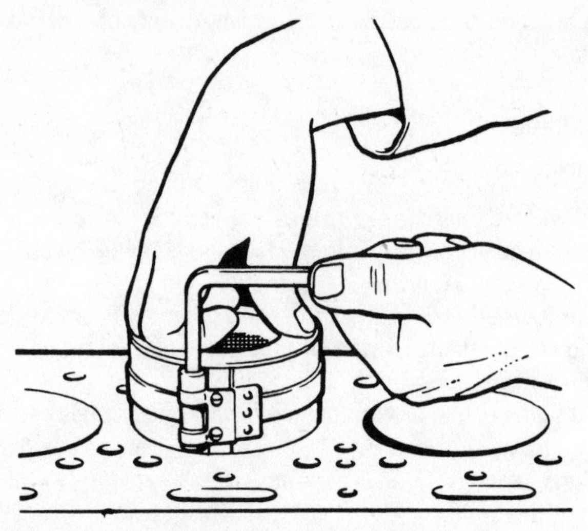

Fig. 7-17. This mechanic is installing pistons the correct way with no more force than can be generated by one hand. (Courtesy Lansing Bagnall, Ltd.)

Reposition the compressor on the piston and try again, this time bearing down on the upper end of the compressor. That will help contain the rings during the transition into the bore.

8. Working now from the underside of the engine, guide the rod over the crankpin. Make certain that the previously installed rod bearing is still in place, and remove the protective hoses.
9. Install the cap with the previously oiled bearing, aligning cap and rod shank numbers. Pull the cap bolts down evenly, and torque in three increments to specification (Fig. 7-18).
10. Rotate the crankshaft. It should turn by hand, dragging a little during midstroke as piston speed (relative to crankshaft movement) increases. You may wish to verify rod bearing clearance with Plastigage.
11. Check big-end side clearance against specification, using a feeler gauge (Fig. 7-19). Inadequate side clearance may mean that the rod cap has

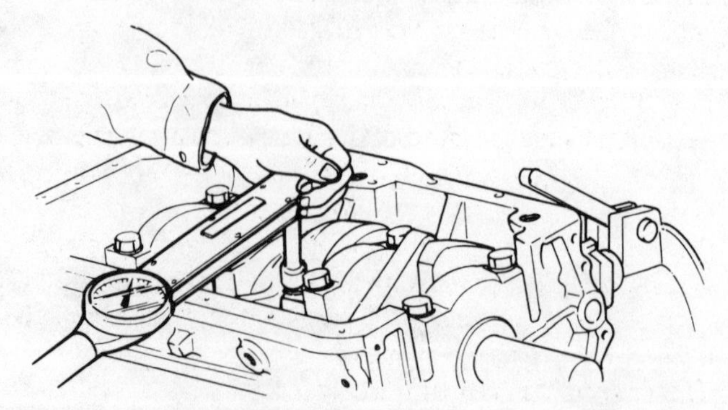

Fig. 7-18. Pull rod caps down evenly in three increments to specified torque.

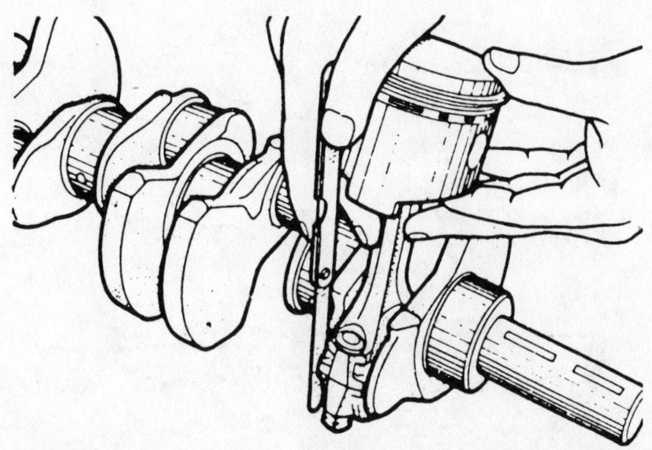

Fig. 7-19. Check big-end side clearance with a feeler gauge. Too little clearance might mean that the cap is reversed or that there is an overgenerous radius on the journal. (Courtesy Chrysler Corp.)

been canted during installation or that the crankshaft radii have been ground incorrectly.

12. Repeat the process for each piston assembly. The crankshaft should be wrench-turnable when the job is completed. However, you may notice increased drag at midstroke, when piston speed increases relative to crankshaft motion.

It is rare to break a ring during installation. But if that should happen, do not purchase another complete ring set until you talk to your machinist. He should have a few loose rings around from accidents in his own shop.

CAMSHAFT AND CAM DRIVE

All premium shops deliver ohv blocks with new cam bearings installed; in most cases, the shop will also prelubricate and install the cam. When cam installation is left to the owner, follow this procedure:

1. Lubricate each journal and each lobe with EP assembly grease rated for camshaft service. (*NOTE:* Motor oil is not the recommended lubricant for either the cam or the lifters.)
2. Mount the engine vertically, nose up, and lower the cam into position, carefully guiding the lobes through the journal bearings and leaving the cam extended a couple of inches out of the bore, so that there is no danger of displacing the welch plug at the back of the block. Lower the engine to the horizontal position and gently push the cam home.
3. Mount the thrust plate, torquing to specification (Fig. 7-20). As noted earlier, on some engines cam "walk" is controlled through the distributor/oil-pump drive gear mesh.
4. The camshaft should turn freely.
5. Check camshaft endplay as described in the appropriate manual.

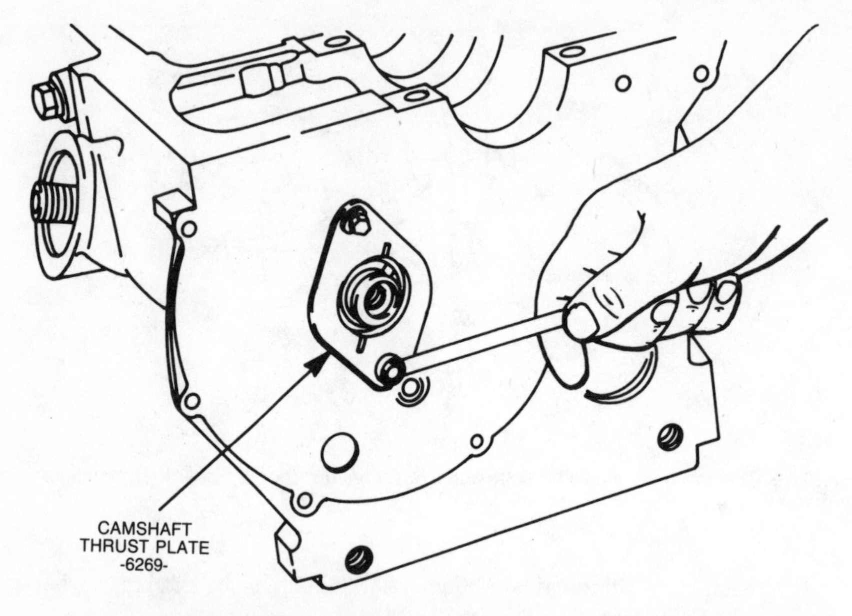

Fig. 7-20. When used, the thrust plate should be lubricated with assembly grease, installed with the rubbing surface adjacent to the camshaft, and torqued to specification. The next step is to determine camshaft end play. (Courtesy Ford Motor Co.)

Install the timing chain and sprockets as an assembly, with timing marks aligned. Most ohv engines call for the timing marks to be aligned on an imaginary centerline through both camshaft and crankshaft bores, as shown in Fig. 7-21. A few ohv engines, such as old Ford Y-block, index colored links on the chain with marks on the sprockets.

It is always a good idea to check gear lash upon assembly; this check is necessary if main bearing saddles have been align bored (Fig. 7-22). This may be accomplished with a dial indicator or by simply rolling a short length of solder through the (steel) gears, on the Plastigage principle. The thickness of the solder represents lash.

Secure the camshaft sprocket, together with other hardware, such as the fuel pump eccentric and crankshaft oil slinger, and pour oil over the chain or gear-drive assembly.

Lubricate valve lifters with the same EP grease used on the cam and install the lifters in their bores. Watch the clearances: as indicated earlier, it is not unusual to find a block with one or more oversized lifter bores.

Mount a new oil seal on the timing cover. Special seal installation tools are available to dealer mechanics; most of the rest of us make do with a length of pipe or dowel slightly smaller than seal OD, so that installation force is confined to the outer, or flanged, edge of the seal. Pry the old seal out with a large screwdriver. Hook the blade under the seal rim and strike sharply on the handle. The seal will pop out. Clean the seal bore, and press or hammer the new seal home. The drive side of the seal—the side that is stressed for installation—will

Fig. 7-21. Chrysler V-8 timing marks are typical of modern ohv practice. Note that the marks are adjacent to each other on a line passes through the crank and camshaft centers.

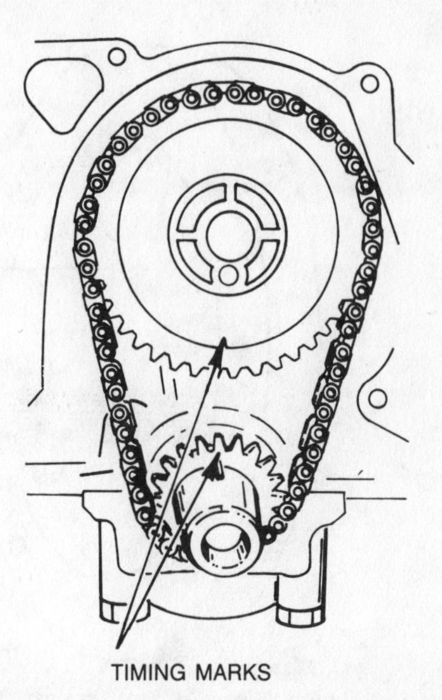

carry the manufacturer's logo and part number. Verify that the steep edge of the lip is inboard to contain oil pressure. Lubricate seal lips with motor oil.

Some front seals ride on the crankshaft, and others on a flanged harmonic balancer. In either case, the seal contact surface must be dead smooth and polished. Fel-Pro makes a sleeve that slips over the harmonic balance shaft to provide a new sealing surface (Fig. 7-23). Victor makes an extended-case oil

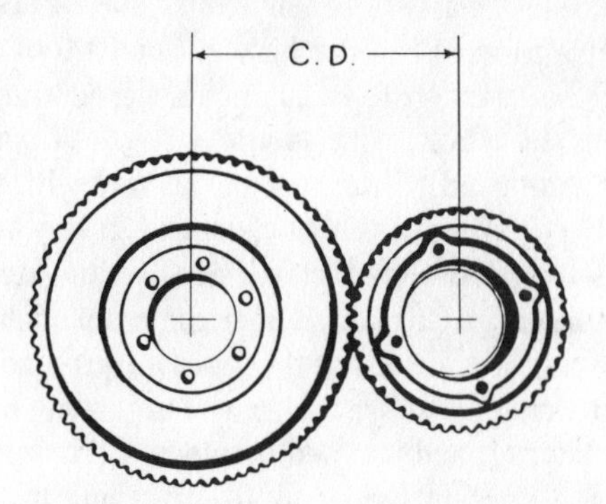

Fig. 7-22. Timing-gear center-to-center distance ("C.D." in the drawing) should be a design parameter, fixed on the drawing board. However, new cam bearings can move the camshaft gear relative to the crankshaft gear; align boring main-bearing saddles will almost certainly dislocate the crankshaft; and new timing gears may introduce inaccuracies of their own. Consequently, gear lash should be checked upon assembly. Because little can be done to correct lash at this late stage in the proceedings, most shops make generous allowance for lash with a minimum of 0.003 in. and a (perhaps noisy) maximum of 0.010 in. (Courtesy Sealed Power Corp.)

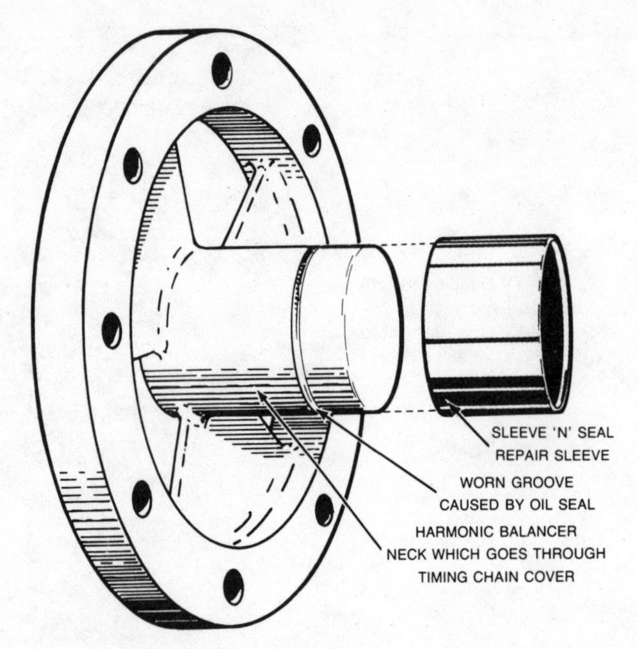

Fig. 7-23. Fel-Pro "Sleeve 'n' Seal" can save the price of a new harmonic balancer.

seal for Chevrolet small-block V-8s that relocates the seal 0.250 in. out on the harmonic balancer shaft, thus presenting a new surface for seal contact. The part number is 65022.

Install the front cover on ohv engines, using a new gasket. Because of the timing requirement, ohc front covers are installed after the head is mounted and before the oil pan.

Pressed-on harmonic balancers benefit from a special installation tool, which is not readily available to weekend mechanics. Such a tool must be used when a press-fit against the crankshaft is all that secures the dampener. Those dampeners that combine a bolt, with an interference fit can be started with a soft-faced hammer and pulled on the rest of the way with the bolt. However, there is a risk that the balancer will be damaged in the process.

Align the keyway, oil the contact surfaces, and—using a lead or brass hammer—drive the balancer far enough over the crankshaft to engage the bolt threads. Tap on the center hub and not the rim. Once threads engage, pull them home with a socket wrench and breaker bar. Torque to specification. A wood block, wrapped with a rag, and jammed between a crank throw and the inner side of the block, will prevent the crank from turning.

FLYWHEEL AND CLUTCH ASSEMBLY

The clutch pilot bearing, which supports the output shaft in the flywheel hub, should be replaced as a matter of prudence. Extract the old bushing or needle bearing and install a new bearing, driving it in from the marked side. This operation is discussed in more detail in Chapter 6.

Flywheel/Flexplate

Install the flywheel or flexplate, aligning the marks made during disassembly. Pull the bolts down evenly, torquing to specification. Remove the wood crankshaft blocker.

Clutch and Pressure Plate

Always install a new clutch disk and throw-out bearing during an engine rebuild. A new pressure plate is optional, but will make clutch action smoother and more predictable. The disk must be centered on the pilot bearing for the transmission to bolt up. Figure 7-24 illustrates an inexpensive alignment tool, available for most popular engines at auto parts houses. The K-D 2420 is a near-universal alignment tool, which includes a tapered shaft adapter and seven pilot-bearing adapters, in both inch and metric sizes. But the best and surest clutch alignment tool is a scrap input shaft.

Follow this procedure:

1. Position the clutch disk (correct side forward) and the pressure plate loosely on the flywheel. Assembly marks must be aligned.
2. Insert the alignment tool through the disk and into the pilot bearing ID.
3. Bolt up the pressure plate, working in the usual crisscross pattern, and tightening bolts no more than 1½ or 2 turns at a pass. Keep the alignment tool dead parallel with the crankshaft centerline.
4. Install the new throw-out bearing, lubricating the yoke bearing surfaces with small amounts of wheel-bearing lube.

When the transmission is out of the vehicle, it is easy enough to check clutch alignment by trial assembly. The input shaft should run into the disk, center on the pilot bearing, and the transmission flange should meet squarely with the bellhousing. If not, loosen the pressure plate and realign the disk.

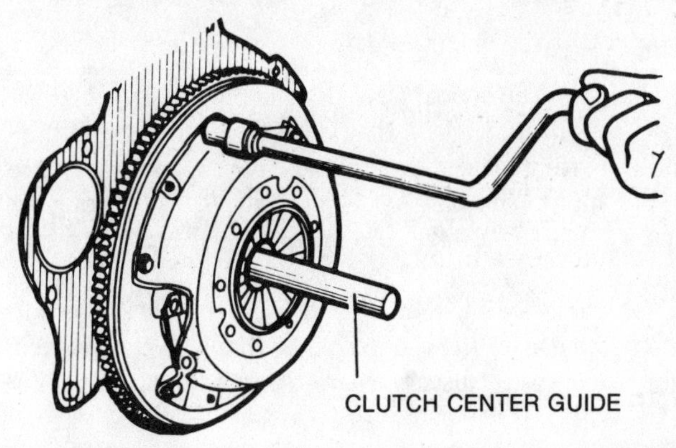

Fig. 7-24. A correctly sized dowel pin can serve as a clutch alignment tool.

OIL PUMP

Submerge the pump in clean oil and prime the unit by rotating the drive shaft. This bit of insurance will protect the bearings during initial start-up.

Some pumps, e.g., pre-1982 small-block Chrysler V-8s, mount into a recess on the block. Misalignment can crack the housing when the hold-down bolts are tightened. Other pumps, such as the GM 173 V-6 unit, secure the driveshaft to the pump with a retainer, that must be engaged for driveshaft retention. The driveshaft is installed with the camshaft in place.

Torque the pump the factory-specified value, pulling the bolts down in a crisscross pattern.

OIL PAN

With the block inverted, set the oil pan in place without a gasket. It should fit flat and square with little or no daylight between bolt holes. Steel pans should be lightly massaged with a ball-peen hammer to correct bolt-hole "pucker." Most aluminum pans cannot—because of the rear main cutout—be resurfaced, but the gasket will compensate for minor deflections. Plastic pans usually will conform under bolt torque.

Beginning in the 1979 model year, Chevrolet 250 and 292 engines six-cylinder truck used a 16-bolt-hole pan, sealed with RTV, rather than a conventional gasket. The experiment ended in 1980 for the larger engine, but the 250 continued to use RTV through 1984. Fel-Pro's PN OS 30462 C is a direct-replacement conventional gasket set.

Torque the fasteners to specification, which will seem to be nowhere near tight enough. But excessive torque will merely distort the gasket.

CYLINDER HEAD

Portions of the cylinder head need attention before your project ends.

Head Bolts

New, one-shot, torque-to-yield head bolts should be supplied with the gasket kit for those late model engines that use such fasteners. If the original bolts can be reused, wire-brush and inspect the threads, and when necessary, chase them with the appropriate die. As always with thread-cutting tools, purchase the best quality that you can find. An undersized die will do more damage than good.

Check for proper bolt length in blind (closed) holes by using a pencil or a drill bit for a depth gauge. If it is determined that a bolt will bottom, put a hardened and machined washer (usually a dealer-only item) under the bolt head before assembly.

Note which, if any, bolts touch water. Head bolt penetration into the water jacket is characteristic of Chevrolet small-block V-8s, some bolts on certain 1979, and all 1980 Chrysler 318s, and many other recent engines. Those bolts that are wetted, and only those bolts, must be coated with a sealant such as Fel-Pro

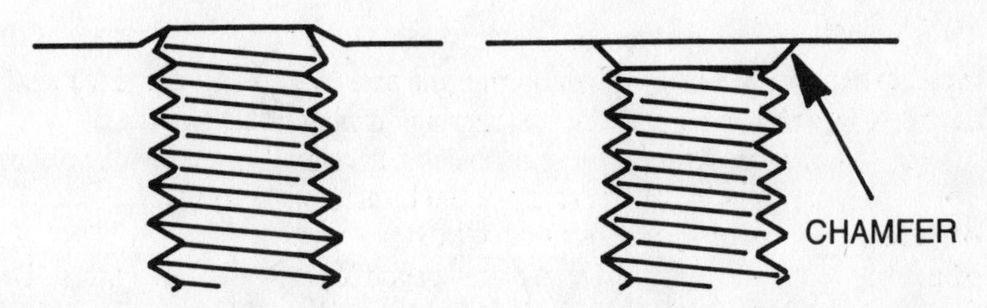

Fig. 7-25. Pulled threads should be relieved as shown with an oversized drill bit, a countersink, or high-speed grinder. (Courtesy Fel-Pro Inc.)

"Pro-Lock 1" or Locktite "Lock-n-Seal." Lightly oil the threads of the remaining bolts.

Inspect bolt holes carefully, and especially if the head or cylinder deck has been resurfaced. Chamfer and retap holes that are "pulled" or threaded flush with the fire deck (Fig. 7-25). A drill bit, ¼ in. or so larger than bolt diameter, will make the required chamfer and chase the threads with a straight-shank bottoming tap. Lacking that—and for some metric threads you will be lucky to find any tap—use a conventional tap with the pointed end ground flat. Clean all debris from the hole. At this point, the block should be dry. But if you have just scrubbed and rinsed the bores or if you are in the midst of a quickie overhaul, water may be lurking in blind bolt holes. Remove with pipe cleaners.

Head Gasket

Position the gasket, oriented with reference to any "Front" or "Top" marks that may be present, on the block deck. Bolt, coolant, and oil holes in the block should align with matching holes in the gasket, although the match is not always perfect, and that is cause for confusion. "Unitized" gaskets, which fit more than one engine model, may have superfluous cutouts. It is also possible for deck and/or head holes to be deliberately blocked.

Production changes have not made things simpler. For example, 1968-73 427 and 1973-79 454 Chevrolet truck engines use a different gasket than later models. Interchanging these gaskets can result in overheating and severe engine damage. The original Buick 231 engine with uneven firing intervals (distinguished by its asymmetrical distributor cam) uses a different gasket than the later models. The catch is that some 1979 even-firing engines were put together with leftover uneven-firing heads and require the earlier gasket. Examples are legion; what is important is to be absolutely certain that the gasket fits the engine at hand. If there is any question about it, check with your machinist or call the gasket manufacturer.

Follow the gasket maker's instructions about the use of adhesives, which are detrimental to some gaskets. You may wish to make a pair of alignment pins from two discarded head bolts. Cut the bolts, just below the heads, and slot the shank for a screwdriver. Loosely thread in the pins on opposite corners of the deck and lower the head into place. Extract the pins and torque the bolts, following the manufacturer's instructions to the letter. Figure 7-26 shows a model

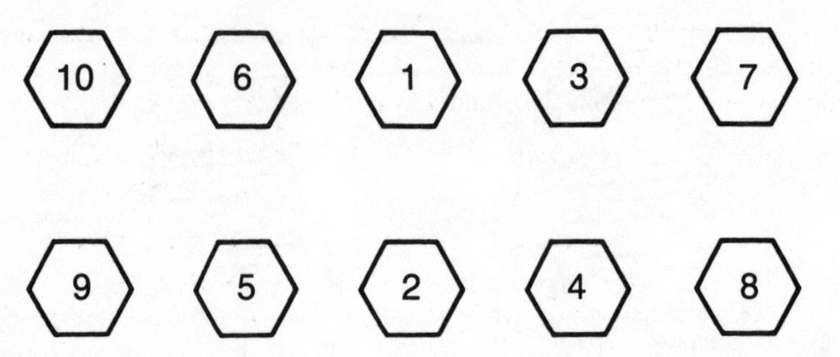

Fig. 7-26. The basic torque sequence is from the center out, so that the head is free to elongate as it is tightened. (Courtesy Fel-Pro Inc.)

torque sequence, but real-world sequences vary from the model. Torque values also may vary between bolts.

Typically, the head bolts are torqued in three circuits of the sequence. First, go around the head, tightening to one-third of final value; repeat the sequence at two-thirds of final value; and make a final pass, this time at full torque. If the bolt makes a popping sound, it has seized and should be loosened and retorqued. This does not hold for torque-turn bolts, which are tightened to a certain ft./lb. value, then turned so many bolt flats past that value.

Most head gaskets do not require retorquing after the engine break-in, although some mechanics take the trouble to do it.

OHV Valve Gear

Install the pushrods, ends well lubricated, in their original positions and orientation. Pour oil over the rocker-arm assemblies, and tighten the hold-down bolts or studs. Some mechanics simply torque the lifters down and leave it at that. But this procedure can bend pushrods. New lifters are fully charged with oil and extended to their maximum lengths. Since oil is virtually incompressible, the lifters act as if they were solid, at least insofar as any rapidly applied force is concerned. None of this would make any difference if all lifters were on the camshaft base circle; but at least half of them will be up on the cam lobes. As the rockers are tightened, the affected valves open, and because the lifters are extended more than normally, the valves may open wide enough to coil bind, locking solid. Something has to give and the pushrods are the weakest links in the system.

Basically there are two ways to cope with this problem, depending upon rocker-arm construction.

Shaft-Type Rockers. When the rockers are arranged on a shaft, like birds on a telephone wire, the procedure is to tighten the shaft hold-down bolts very slowly, allowing a minute or more for lifter bleed-down. Do not rotate the engine before the lifters have time to retract.

Pedestal-Type Rockers. All lifters will be on the camshaft base circle during the course of a single crankshaft revolution from tdc. For example, the sequence for Ford 352, 390, and 428 engines is as follows:

Crankshaft Position	*Lifters on Base Circle*
Tdc	Intake: 1,3,7,8 Exhaust 1,4,5,8

Three hundred sixty degrees later, the remaining lifters are zeroed. The sequence varies with firing order, but can be determined from the shop manual or from observation.

OHC Valve Gear

Install the timing chain or belt, aligning the timing marks and tensioning the belt with religious adherence to factory instructions. Lubricate the camshaft, rocker arms, and timing chain as described above. Pull the engine through several complete revolutions by hand to verify that pistons and valves clear. Recheck the timing and install the front cover.

Initial Valve Lash Adjustment

Nonadjustable rockers (e.g., Chrysler small-block V-8s) are torqued down to specification; adjustable pedestal-type rockers are tightened against the base circle (until the pushrod can no longer be turned by hand), then preloaded the prescribed number of turns. For example, most GM rockers are tightened one or two turns beyond pushrod lock. This positions the lifter pistons near the mid-range of their 0.150-in. travel. US-built Ford engines use three pushrod lengths—standard, 0.060-in. under, and 0.060-in. over—to obtain correct plunger travel. Special tools are required to collapse the lifters and determine correct pushrod length. However, this should not be required if the pushrods are installed as originally found and if the machinist has held tolerances.

Mechanical, or solid, lifters include a provision for adjustment at the rocker arms. (Certain sidevalve engines, now obsolete, were adjusted by grinding the valve stems.) The manufacturer may provide both "Hot" and "Cold" specifications, or merely the "Hot" spec, which is valid only after the engine has run for 20 minutes or more. In that case, adjust the valves for 0.005 in. more lash than the "Hot" specification.

INTAKE MANIFOLD

The "trick" in mounting the intake manifold is to maintain gasket integrity, especially on vee-type engines where the gasket must seal air, water, and oil. The following procedure, abstracted from Fel-Pro recommendations, will do much to eliminate problems:

1. Clean all mounting bolt threads.
2. Check for manifold fit with a straightedge and feeler gauge. In-line manifolds can be easily resurfaced; severaly warped vee-type manifolds might have to be replaced.
3. Check gasket fit.

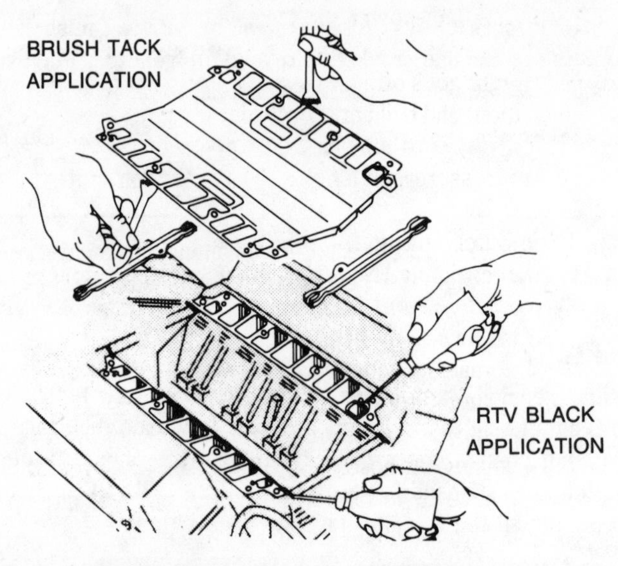

Fig. 7-27. Intake manifold gasket installation on a V-8 engine. This particular engine uses a type of gasket known as a valley pan (also called a bathtub, turtleback, or turtlepan) and strip seals. (Courtesy Fel-Pro Inc.)

4. Steel shim gaskets usually require application of adhesive around intake openings and a light coat of RTV at vacuum ports. Follow the instructions packaged with the set.
5. Apply a flexible sealant to bolt threads that penetrate water passages.
6. Use an appropriate adhesive, such as "Brush Tack," to secure the gasket on vee-type engines and seal the ends of strip seals with RTV (Fig. 7-27). Strip seals on some Chrysler engines are deliberately cut ¼ in. short. The resulting gap must be filled with RTV.
7. Follow the bolt-tightening OEM or gasket maker's torque sequence. Torquing is usually begun at the center of the manifold, working outward in an X-pattern (from bank to bank on vee-type engines).

FINAL ASSEMBLY

Fill the sump with 30W SF motor oil, install a new filter (Consumer Reports recommends Fram), and if at all possible, protect your investment by pressurizing the oiling system. Although bearings were prelubed, and may survive because of it, oil pressure will be erratic until the pump generates enough volume to purge air for the circuits and fill the main gallery and filter canister.

It is sometimes possible to drive the oil pump with an electric drill motor and a long screwdriver bit. Turn the engine in the normal direction to determine pump driveshaft rotation and power the pump until a steady oil flow appears at the (non-pushrod-lubricated) rockers or, if the engine is in the car, until the oil-pressure warning light goes off. An alternate method is to remove the spark plugs, disable the ignition, and build pressure with the starter motor. The surest and most professional way is to pressurize the system with an external pump, connected to the oil pressure sender line. The pump can be used to fill the crankcase.

Install the distributor, using the marks made previously to obtain an approximation of correct timing. Note that the distributor/oil pump driveshaft may have a preferred orientation. Snug the distributor clinch bolt just enough to allow the unit to be turned by hand.

Assemble the remaining components, double-checking all plumbing connections and using new filters.

Lower the engine into the bay and into engagement with the transmission input shaft. Watch for dangling hoses and wires. Both the back of the engine and transmission may have to be raised on jacks. A pair of guide pins (similar to those described under "Cylinder Heads") simplifies alignment. If the input shaft binds against the clutch disk, put the transmission into gear and rotate the crankshaft a few degrees. Once the transmission is "stabbed," remove the guide pins, bolt up transmission, and secure the motor mounts. Replenish any oil that was lost from the transmission, and make the remaining under-vehicle connections.

Complete the topside installation, top off the radiator, and install a fully charged battery. The engine will start easier if the carburetor float bowl is filled with fuel. Verify that dashboard oil pressure, engine temperature, and charge indicators function.

START-UP AND RUN-IN

Make one last check of fuel-line connections and liquid levels. Connect a timing light, and stand by the engine compartment while a helper cranks the engine and monitors dash instruments. It may be necessary to turn the distributor a few degrees to coax the engine into life. Once the engine starts, set the ignition timing to specs, tighten the distributor hold-down bolt, install the air cleaner, and run the engine to a fast idle—about 1500 rpm. If the coolant level drops, shut the engine off and let it cool before adding cold water. Restart and continue to run as before for 30 minutes. Check for leaks, above and below, continue to monitor dash instruments, and adjust the carburetor as necessary.

Adjust the valve lash on engines with mechanical or adjustable hydraulic lifters, either with the engine ticking over at a slow idle or by the method described under "Initial Valve Lash Adjustment" above. Figure 7-28 illustrates the three-handed adjustment process for mechanical lifters: loosen the locknut and insert the appropriate feeler gauge blade between the valve stem and rocker tip. The gauge will bind and release with the rise and fall of the lifter. Light drag signals that lash equals blade thickness and that the adjustment is correct. Tighten the locknut and recheck the clearance.

Fig. 7-28. A special tool that combines a screwdriver with a box-end wrench makes it easier to adjust mechanical lifters. But, with a little patience, one can adjust the lifters as shown. (Courtesy Lancing Bagnall, Ltd.)

Running adjustments for hydraulic lifters are simpler: loosen the adjuster until the rocker rattles, wait several seconds for the lifter to pump up, and slowly tighten the adjuster until the noise stops. This is 0 lash. Screw down the adjustment the specified number of turns to center the lifter piston in its bore. The tendency of pushrod-oiled rockers to act like lawn sprinklers when the valve cover is removed can be controlled by blanking off the rocker oil ports, with special plugs, short lengths of rubber hose, or even ordinary clothespins.

Make a test drive at 30 mph, accelerating rapidly to 50 mph and decelerating. Repeat this procedure—which, as brutal as it sounds, helps seat the rings—at least ten times. Drive conservatively for the first 500 miles, with occasional bursts of acceleration. Change the oil and filter. After 1000 miles, check timing and adjust valve lash one more time.

Congratulations!

Index

Index

D

E

R

S

T

U